John Silve[illegible]

Forest Leaves

Mr. S,
This was a great year with you and I enjoyed all our projects.
Thank you
Cooper 1997

Forest Leaves

How to Identify Trees and Shrubs of Northern New England

❖

by Henry Ives Baldwin

with original drawings by
Gunnar Ives Baldwin, Jr. & Priscilla Kunhardt

PETER E. RANDALL PUBLISHER
Portsmouth, New Hampshire

This book was originally published by the Society for the Protection of New Hampshire Forests, New Hampshire's oldest and largest nonprofit conservation organization. The Society encourages broader understanding of our unique forest heritage as a foundation for careful natural resource stewardship. The Society, based in Concord, New Hampshire, is proud to have counted Henry Baldwin among its most ardent supporters—as trustee, educator, author, and friend.

Printed in the United States of America
Published 1982. Second Edition 1993

Designed by Debra Kam
Cover photograph by Peter E. Randall
Photograph of the author courtesy of the Baldwin family

Peter E. Randall Publisher
P.O. Box 4726
Portsmouth, NH 03802-4726

Library of Congress Cataloging-in-Publication Data

Baldwin, Henry Ives, 1896–1992

Forest leaves: how to identify trees and shrubs of northern New England / by Henry Ives Baldwin; with original drawings by Gunnar Ives Baldwin, Jr. & Priscilla Kunhardt.

p. cm.

Includes bibliographical references (p.) and index.

ISBN 0–914339–43–5: $12.95

1. Trees--New England--Identification. 2. Shrubs--New England--Identification. 3. Leaves--New England--Identification. 4. Forest plants--New England--Identification. I. Title.

QK121.B34 1993

582. 160974--dc20 93-2505

CIP

Contents

Preface

Leaves

Leaves are not the only way to identify trees and shrubs, or not always the best way. However, leaves (and needles) are always present in summer and sometimes in winter. Their shape, color, texture and odor often serve to identify the plant by a leaf alone. Accordingly, leaves are illustrated in this book, and other features described. That is why this book differs from others. Leaves can be collected easily and pressed between the pages of a magazine for later study indoors. Leaves can be collected from the ground beneath a tree where leaves cannot be reached on the tree.

Very often the leaf alone will suffice for identification.

With few exceptions, the leaves illustrated are natural size. Other features of diagnostic value are given in columns in the text. More important clues to identity are shown in capitals, less critical, but still valuable features in italics, and supporting information in roman type.

The Drawings

I am greatly indebted to the two fine artists who made this book possible; and generously gave their time and talents.

The drawings of leaves and other plant parts are the work of Gunnar Ives Baldwin, Jr. (106 drawings) and Priscilla Kunhardt (20 drawings). Drawings were sketched freehand from fresh or pressed leaf specimens. The original drawings were all life-size. Some of the larger leaves are reduced. It should be pointed out that leaves of any species, or even those on the same tree, vary greatly in size.

Acknowledgments

Special thanks are due Patricia Churchill of the Society for the Protection of New Hampshire Forests who did all the typing and collation of the first edition. I am indebted to the late Lesley Clark for much assistance, and to Brian Simm for help with the second edition.

Introduction

The purpose of this book is to provide brief, simple and easy identification of common trees and shrubs in northern New England. Many uncommon species are omitted. Technical terms are avoided.

Why is such a book needed? There are many excellent manuals that cover these plants. Every state has its own manual. Nearly all these books are excellent and should be consulted to supplement this outline and to identify plants not included. The reason for this book is that (1) most tree books devote a long paragraph to describe each species. The reader has to search for the important identifying features; (2) these unique features need to be given emphasis; often one character alone serves to identify a tree in this part of the country. Example: white pine is the only 5-needled pine; pitch pine is the only native 3-needled pine. Here, instead of a long essay, each species is listed with its characteristics in outline form. These aim to be complete. The reader can skip the detail if he can finalize the identity by one or two words. However, to be sure, during all seasons there are other clues.

What is a tree and what is a shrub?

They are both woody plants, usually perennial (that is living more than one growing season). How are they distinguished?

Trees may be defined as woody plants that:

(a.) Usually grow up from the ground with a single erect stem or trunk. This main stem may be massive; it is often unbranched for several feet above the ground.

(b.) Form a more or less definite pattern of crown or branch arrangement. The habit and form of growth may be more important than its height. This applies especially to young trees.

(c.) Reach a considerable height at maturity, at least 8 to 20' (2–6 m), and with a diameter 2 to 3" (5–8 cm) at breast height. Authorities differ.

When is a tree not a tree? When it is a shrub!

Shrubs are (generally) considered to be shorter than trees and smaller in trunk size. Shrubs are usually characterized by:

(a) a cluster of stems rising directly from the ground. There are almost invariably several stems, erect, spreading or prostrate.

(b) a general bushy appearance, with no special crown shape.

(c) rarely exceeding 12 to 20' (3–6 m) in height at maturity.

There are always what appear to be exceptions, and no hard and fast line can be drawn between a tree and a shrub. For example, tree seedlings 3 to 5' (1–2 m) high, such as birch or maple, may grow so densely that they look like "a cluster of stems arising directly from the ground." Root suckers from beech and other hardwoods exhibit this appearance. Sprouts from low stumps look like shrubs. A single shoot of a shrub resembles an immature tree. How can one be sure if the shoot is mature or not? Identification of the plant itself is more important than deciding to which category it belongs. That is the purpose of this outline.

How to identify trees and shrubs in the field

Plants can be identified by photographs, drawings and descriptions. Some can be told by feel, taste, smell, color and form. The habitat or place where they grow can be helpful. Drawings are often better than photographs to show detail. One can also compare a plant with herbarium specimens, collections of woods, barks, seeds, prints of leaves, etc. Or, one can "run down" a plant in an analytical key. Some short cuts are given at the end.

Most trees have characteristic profiles, stem form and crown shape. However, these vary with the site where the plant grows. Trees growing in the open are broad, spreading, quite different from their natural form in a dense forest. Few profiles are sure clues to identification.

Different factors can be used at different seasons. Buds in winter are very useful in hardwoods (although the same buds may be present at other seasons). Autumn foliage differs among plants and can be of diagnostic value.

The smell or taste of broken twigs, crushed foliage or bark can be a good hint. The sticky gum of buds, resinous smell, and the sharpness of needle points, spikes and thorns on leaves and twigs are all important.

Some trees like tamarack, Atlantic white cedar and black spruce grow in swamps and bogs, red cedar and juniper never; the latter live on dry pastures.

Branching habit is most important. Opposite branching is where leaves and branches form opposite one another in regular whorls as in most maples. Alternate or irregular branching is characteristic of hemlock and many hardwoods.

Leaf form is perhaps the best help in summer with broad-leaved trees, and at any time with evergreens. Simple leaves are just one blade on a stalk; compound leaves like ash, hickory, butternut and mountain ash have two or more leaflets from a single leaf stem. Leaf margins may be smooth (entire) or with various kinds of teeth, lobes, spikes and humps. The pattern of veins in the leaf is often distinctive.

The species described here are arranged according to botanical order. Needle-leaved trees come first.

How to use this book

Usually one or two characters will serve to identify the tree or shrub. The key features are in CAPITALS, the next most important in *italics*. So why bother with all the other features? Simply because someone might get interested and want to identify the plant in winter, or, what if the key features are missing?

There are often many common names in use for the same tree or shrub. The preferred common name is given first, followed by the most often used synonyms in ( ). The scientific name follows. *Gray's Manual of Botany, 8th Edition*, is used for both common and scientific names. Metric equivalents are included.

The designation "*N.T.S.*" indicates drawings that are scaled not to size.

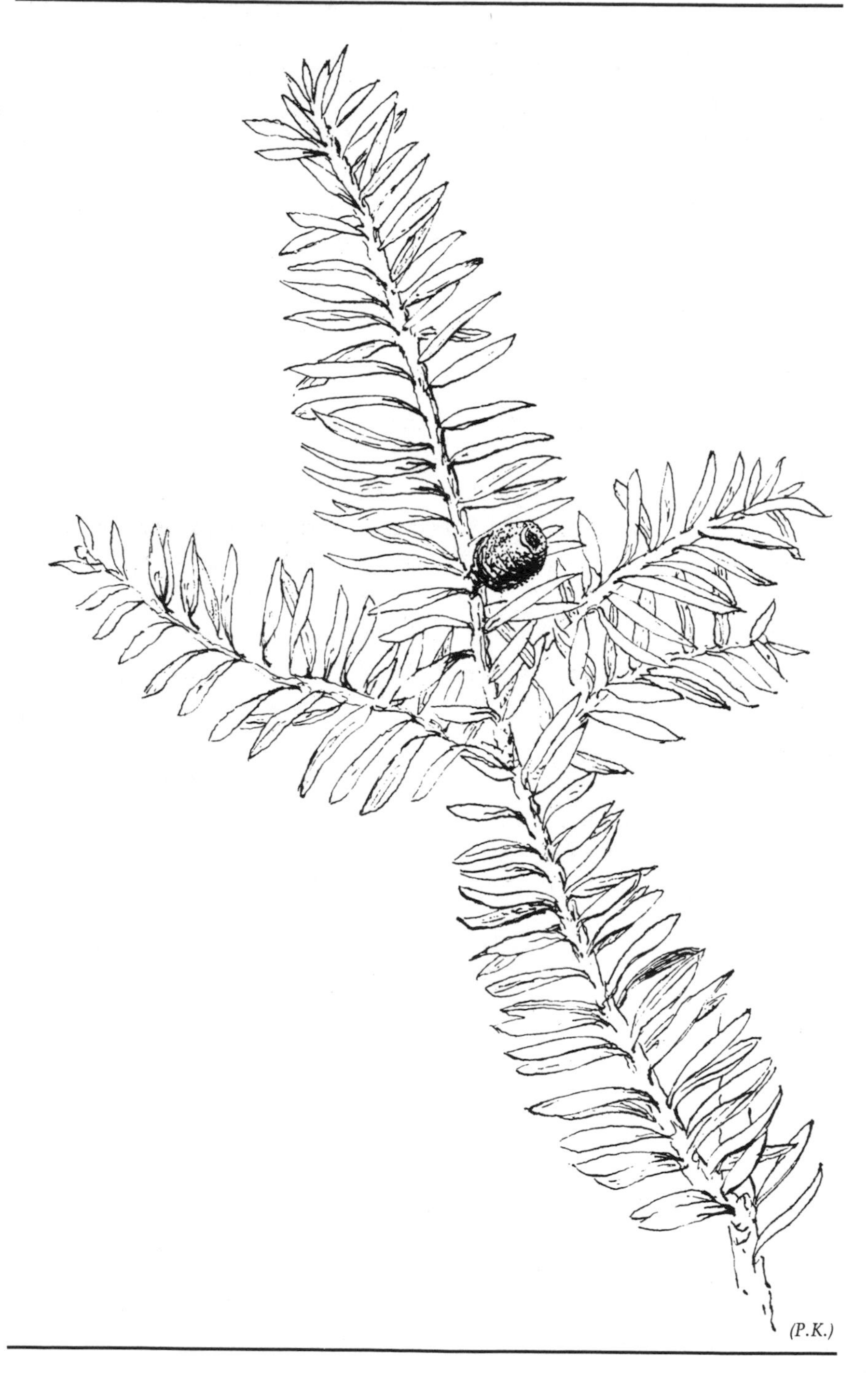

American Yew

American Yew

(Ground Hemlock)

Taxus canadensis

Range: Northeast U.S. Infrequent in the north.

Habitat: Marsh.

Profile: Small, evergreen shrub.

Branching: Alternate or irregular.

Leading Shoot: Drooping or indeterminate.

Leaves: Narrow, *fine-pointed*, dark yellow-green, *stalked*. UNDER-SURFACE ENTIRELY GREEN.

Buds: Scales long and narrow.

Twigs: BRIGHT GREEN becoming brown on older twigs.

Bole: Creeping with ascending or upturned branches.

Bark: Rough, scaly.

Fruit: A RED BERRY, edible, but seeds deadly poisonous.

(G.I.B.)

Red Spruce

Red Spruce

Picea rubens Sarg.

Range: The Northeast and Appalachian Mountains to Georgia.

Habitat: Upland; mostly higher elevations.

Profile: Rounded crown.

Branching: Opposite.

Leading Shoot: Straight, rarely weevilled.

Leaves: Needles short 1/2–3/4" (12–19 mm). Dark or yellow green, *no "bloom."* Needles crowded, incurved, *twisted* mostly on *upper side* of twig. *Needle stalks prominent.*

Buds: Oval, pointed, *reddish* brown.

Twigs: VERY HAIRY, REDDISH BROWN, usually upcurving.

Branches: Horizontal, coarse.

Bole: Moderate taper.

Bark: Thin, scaly, REDDISH brown.

Flowers: Male: *bright* red. Female: greenish red to pink. Both on same tree.

Fruit: Immature: cone, nearly ROUND or *egg-shaped, green* or *purple*, 1–2" (3–5 cm). SCALES SMOOTH. Ripe: brown or reddish brown.

(G.I.B.)

White Spruce

White Spruce

(Cat Spruce, Skunk Spruce)

Picea glauca (Moench.) Voss.

Range: North of White Mountains.

Habitat: Lowland. Moist sandy loams, or alluvial soils; especially typical of stream borders, lake shores and adjacent slopes.

Profile: Columnar crown in older trees.

Branching: Opposite.

Leading Shoot: Straight.

Leaves: *Longer*, 1/2–1" (12–25 cm), BLUISH GREEN, SLIGHT BLOOM, PUNGENT ODOR WHEN CRUSHED, sharp line of white dots in each groove. *Needle stalks inconspicuous.*

Buds: Oval, ragged, *brown*, smooth.

Twigs: PALE GRAY to YELLOW brown, NO HAIRS OR FUZZ.

Branches: Horizontal, often very coarse.

Bole: Moderate taper.

Bark: Scaly, GRAYISH brown, *silvery on freshly exposed areas.*

Flowers: Male: *yellowish* red. Female: greenish red, both on same tree.

Fruit: CYLINDRICAL 2" (5 cm) long. Immature: THIN SCALES, SOFT AND FLEXIBLE. Ripe: *light chestnut brown.*

(G.I.B.)

Black Spruce

Black Spruce

(Bog Spruce)

Picea mariana (Mill.) Bsp

Range: Eastern U. S., Northern Canada.

Habitat: *Bogs* and tundra.

Profile: *Sharp-pointed, narrow crown*, crowded cone area at top.

Branching: Opposite, drooping.

Leading Shoot: Straight, never weevilled.

Leaves: VERY SHORT needles 1/4–1/2" (6–10 mm), pale BLUISH GREEN with *whitish bloom.* 4-sided, *blunt-pointed* white lines broader above than below.

Buds: Sharp pointed, *lowest scales fuzzy* with a ring of pointed scales at base.

Twigs: HAIRY, dark brown to black.

Branches: Slender, often drooping in old trees.

Bole: Slender, with little taper.

Bark: THIN, SCALY, GRAYISH black. Freshly exposed *inner layers olive green.*

Flowers: Small separate cones on same tree. Male: bright red; female: purple.

Fruit: Egg-shaped to round cones 1–1 1/2" (2.5–3.8 cm) gray-brown, PERSISTENT MANY YEARS. Cone scales, *brittle, toothed on margin.* Purple when young.

(G.I.B.)

Norway Spruce

Norway Spruce

Picea abies (L.) Karst.

Range: Originally central Europe; recently introduced.

Habitat: *Loamy soils*, hills and mountains.

Profile: *Broad crown.*

Branching: Opposite, variable, drooping.

Leading Shoot: Straight, *often weevilled in U.S.*

Leaves: LONG, DARK YELLOW-GREEN, *glossy, stiff and sharp, 2–3 faint lines* of white dots on each groove, beneath. *Needle stalks prominent, rounded.*

Buds: Conical, pointed, non-resinous, 1/4" (6 mm), light brown.

Twigs: REDDISH BROWN, *smooth,* no hairs.

Branches: Stout, horizontal or drooping, rarely ascending, often comblike.

Bole: Stout, moderate taper.

Bark: REDDISH BROWN with thin surface scales.

Flowers: Small separate cones on same tree. Male: bright red; female: purple.

Fruit: VERY LARGE, CYLINDRICAL, 4–8" (10–20 cm) long purple to green when young; chestnut brown when ripe; drooping when ripe.

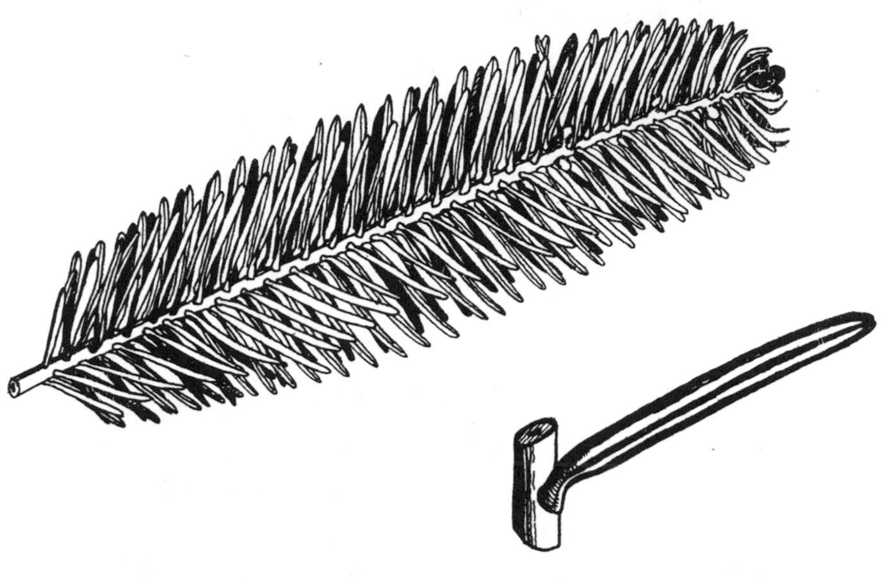

Balsam Fir (G.I.B.)

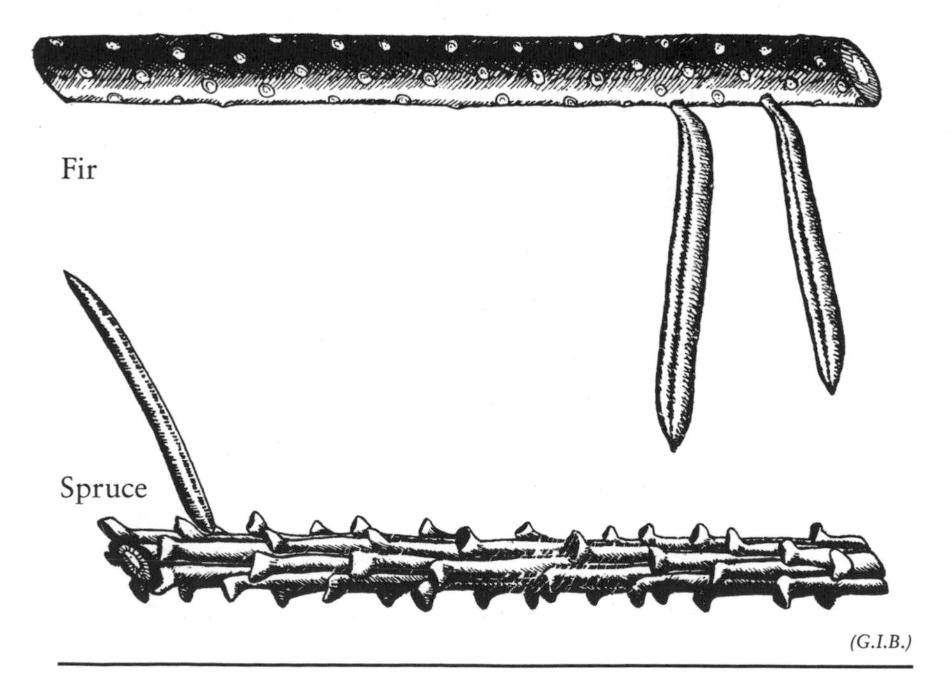

Comparison of Fir and Spruce Twigs

Balsam Fir

(Fir Balsam)

Abies balsomea (L.) Mill.

Range: Common in north.

Habitat: Restricted to moist, shady areas in south.

Profile: *Narrow spire* when mature.

Branching: OPPOSITE, *whorled.*

Leading Shoot: Erect, straight.

Leaves: Flat, 1/2–1 1/2" (12–38 mm) LONG, dark green, *lustrous* on top, 2 white lines beneath. NOTCHED AT END. NO STALKS, LEAVING A CIRCULAR SCAR ON TWIG. FRAGRANT WHEN CRUSHED.

Buds: 1/8–1/4" (3–6 mm), orange-green scales. VERY RESINOUS.

Twigs: *Smooth,* soft, grayish. Needle scars show as DEPRESSED WHITE DOTS. OPPOSITE branching.

Branches: *Horizontal, straight.*

Bole: *Slender* with little taper.

Bark: Ash-colored, *smooth on old* trees. With *resin-filled blister.*

Flowers: Male: yellow and purple. Female: purple scales. Both on same tree.

Fruit:, Immature: Cylindrical cone 1 1/2–3" (38–76 mm). ERECT ON UPPER SIDES OF BRANCHES, falling apart at maturity, leaving a central *spike.*

(G.I.B.)

Eastern Hemlock

Eastern Hemlock

Tsuga canadensis (L.) Carr.

Range: Common everywhere except high mountains and far north.

Habitat: Many types of soils. It's best development is in cool, moist situations.

Profile: *Broad pyramid.*

Branching: ALTERNATE, or *irregularly opposite.*

Leading Shoot: DROOPING, NODDING, NOT ERECT.

Leaves: Needles flat 1/4–2/3" (6–15 mm). SHORT, dark, yellow-green, *tapering,* 2 white lines beneath, appearing as a broad stripe. *Stalked with base persistent comb-like arrangement on twig. Flattened.*

Buds: 1/2–3/4" (12–19 mm) thick, oval chestnut brown, *hairy scales.*

Twigs: Yellow, grayish brown, *very hairy, flattened.* ALTERNATE BRANCHING. *Secondary branches 3–4 times* oppositely divided.

Branches: *Often drooping.*

Bole: Often slightly *tapering.*

Bark: Cinnamon red to gray, tinged with *purple,* narrow ridges.

Flowers: Male: yellow. Female: pink or pale green. Both sexes on same tree.

Fruit: Immature: Oval cones, 1/2–3/4" (12–19 mm). SMALL ON SLENDER STALKS. Cone scales as wide as long.

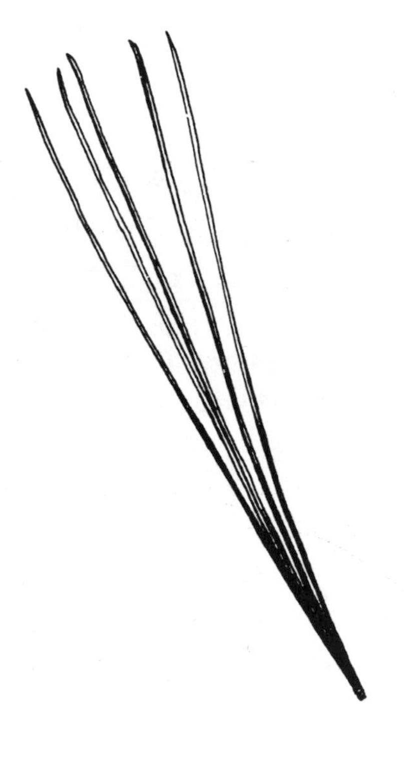

(G.I.B.)

White Pine

White Pine

(Eastern White, Weymouth)

Pinus strobus L.

Range: Northeast U.S. and southern Canada.

Habitat: Lower elevations, sandy soils.

Profile: Rounded crown, *often forked* because of weevil damage.

Branching: Opposite, in whorls.

Leading Shoot: Often killed back by weevil.

Leaves: IN CLUSTERS OF 5, *soft, slender,* often bluish green. Retained to the end of the second growing season. 3-year needles yellow in October. 2–5" (5–12 cm).

Buds: Sharp-pointed, yellowish brown.

Twigs: Smooth, slender, *green or light yellowish green,* resinous.

Bole: Straight, moderate taper.

Bark: *Smooth, greenish when young,* becoming dark brown and furrowed in older trees.

Flowers: Male and female as separate small cones on the same tree. Male flowers oval 3–4' (8–10 cm).

Fruit: *Large cylindrical* cones 4–8" (10–20 cm), *stalked,* long tapering, falling when ripe (August–September). Requires 2 years to mature.

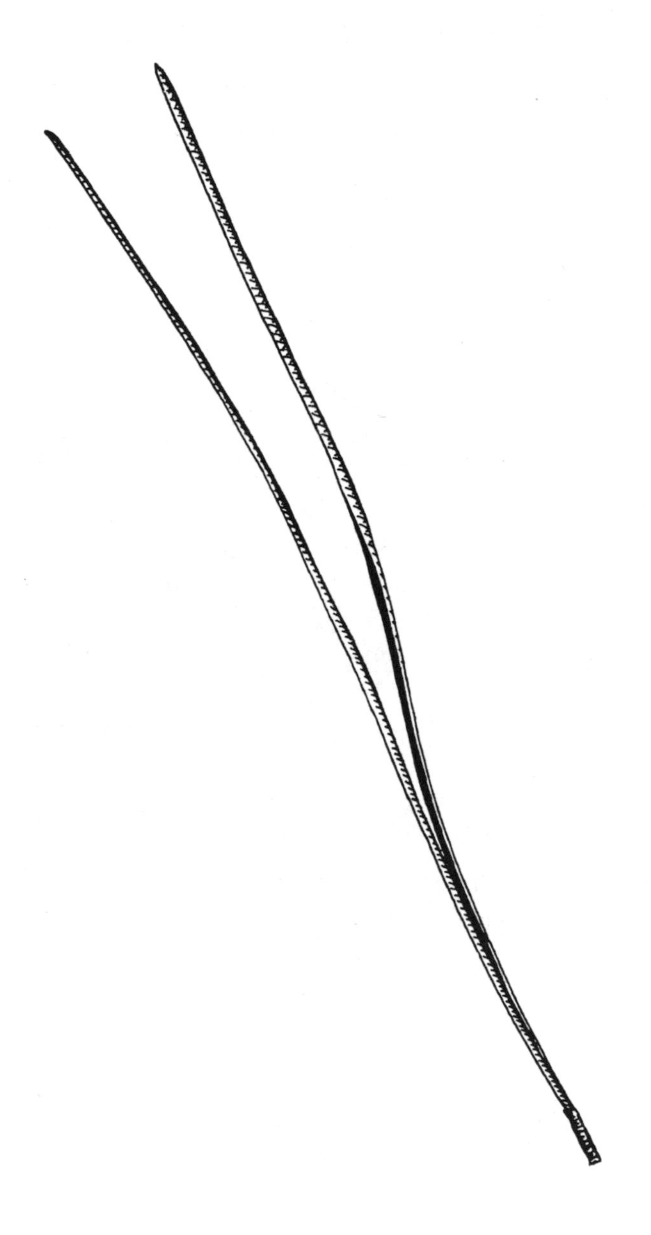

(G.I.B.)

Red or Norway Pine

Red or Norway Pine

Pinus resinosa Ait.

Range: Northern U.S. and southern Canada.

Habitat: Dry, gravel and sandy sites.

Profile: Rounded crown.

Branching: Opposite in two's.

Leading Shoot: Straight.

Leaves: IN CLUSTERS OF 2, stiff, straight, LONG 4–7" (10–18 cm), dark green. *Break cleanly when bent between fingers.*

Buds: Light reddish brown, resinous.

Twigs: Stout, *rough,* reddish brown.

Bole: Straight, little taper.

Bark: *Rough, scaly, reddish,* even on young trees.

Flowers: Male and female as separate small cones on the same tree. Male flowers 5–7" (12–18 cm).

Fruit: *Small, oval* 2" (5 cm), sealed *directly on twig* (no stalk) falling late, within a year. Require two seasons to mature.

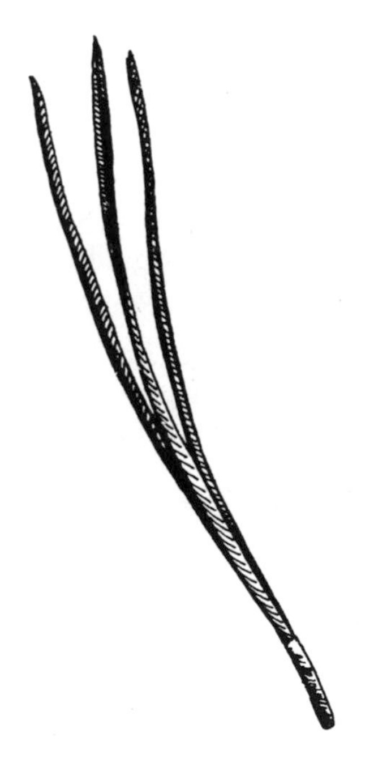

(G.I.B.)

Pitch Pine

Pitch Pine

Pinus rigida Mill.

Range: Eastern U.S.

Habitat: Dry sandy "pine barrens."

Profile: Rounded crown, broad.

Branching: Opposite.

Leading Shoot: Straight.

Leaves: IN GROUPS OF 3 (only native pine in new England with 3 needles) 1–4" (2.5–10 cm), long, yellow, green, stiff, mostly dropping at end of second season. Arising at right angles to branch, *spreading when mature.* OFTEN SPROUTING FROM BARK AND TWIGS.

Buds: Pointed, resinous, chestnut brown.

Twigs: Stout, light brown.

Bole: Short, massive, tapering.

Bark: Thick and rough even when young, dark and furrowed. CLUSTERS OF NEEDLES GROWING OUT OF BARK.

Flowers: Male and female on same tree.

Fruit: Small oval cones 1–3" (2.5–7.6 cm). OFTEN CLUSTERED, STALKLESS, scales with prickles. PERSISTENT ON THE TREE MANY YEARS.

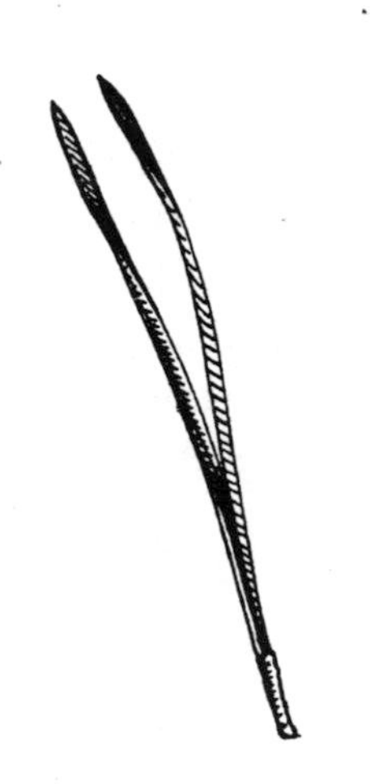

(G.I.B.)

Scotch Pine

Scotch Pine

(Scot's Pine)

Pinus sylvestris L. (exotic)

Range: Western Europe, recently introduced.

Habitat: Plains, especially sandy areas.

Profile: Narrow, pointed crown in north; commonly broad crown in U. S.

Branching: Opposite.

Leading Shoot: Often bent.

Leaves: Needles in groups of 2. Short 1½–3" (4–7 cm), *stiff, twisted, yellow-green to blue-green* depending on seed source and season.

Buds: Short, pointed, light brown, very resinous.

Twigs: *Rough, dull* gray to *yellow,* often reddish or purple brown horizontal to ascending.

Bole: OFTEN CROOKED IN U. S.

Bark: Rough graying brown, scaly, UPPER PART OF BOLE OFTEN PAPERY, REDDISH ORANGE.

Flowers: Male: yellow-orange in dense clusters. Female: reddish on end of shoot.

Fruit: Cones oval about same size as pitch pine, ON SHORT STALKS, FALLING WHEN MATURE. Prickles minute or wanting.

(G.I.B.)

Jack Pine

Jack Pine

(Gray, Scrub or Banks Pine)

Pinus banksiana Lamb.

Range: Extreme northern U. S. and Canada. Extremely rare in New England.

Habitat: Dry, sterile sand and ledges.

Profile: Crown pointed and slender in youth often becoming irregular and rounded in older trees.

Branching: Opposite.

Leading Shoot: Erect.

Leaves: *In groups of* 2. Thick, gray to yellow green, CURVED AND TWISTED AT BASE, DIVERGING AT A WIDE ANGLE. *Very short* 3/4–1 1/2" (2–4 cm), persist 2–3 years.

Buds: Short, pointed, light brown, very resinous.

Twigs: Slender, reddish, often purplish brown.

Bole: Variable from straight to very wavy and crooked.

Bark: Dark reddish brown, scaly.

Flowers: Male and female on same tree. Male flowers densely massed at end of twigs, yellow-brown. Female minute, reddish at twig ends.

Fruit: Conical, 1 1/2–2" (4–5 cm) long. CURVED, OFTEN POINTING TOWARD TWIG, scales with *minute prickles. Persist on the tree for many years.*

N.T.S. (G.I.B.)

Tamarack

Tamarack

(American Larch, Eastern Larch)

Larix laricina (DuRoi) K. Koch.

Range: Eastern U.S. and Canada.

Habitat: Chiefly bogs and Arctic tundra.

Profile: Pointed top.

Branching: Opposite, short, horizontal branches.

Leading Shoot: Long, straight.

Leaves: Deciduous needles, long flexible *triangular, rounded above, keeled* on lower surface. *Bright bluish green,* no white bands 3/4–1 1/2" (1.9–3.8 cm).

Buds: Dark, reddish brown, globe-like, RESINOUS.

Twigs: SMOOTH, BROWNISH, MANY leaf scars.

Branches: Slender.

Bole: Slender.

Bark: *Reddish brown, scaly.*

Flowers: *Female: rose red* with elongated green tips. Male: yellow.

Fruit: SMALL CONES 1/2–3/4" (1.2–2 cm) with 20 scales. Erect. Cone scales CONCAVE, *smooth often shining, longer than broad, chestnut brown, smooth on outside.*

Remarks: *Slow-growing,* mostly short, with much *taper in maturity.*

N.T.S. (G.I.B.)

European Larch

European Larch

Larix decidua Mill.

Range: Central Europe: Italy to Poland. Recently introduced.

Habitat: Upland to timberline.

Profile: Pointed top.

Branching: Opposite, long, often ascending.

Leading Shoot: Straight, *often very long* (1 m).

Leaves: Flat, keeled *below, bright yellow* green, faint or no white bands. 1 1/2" (3.8 cm) longer on short growth.

Buds: Golden brown. NO RESIN.

Twigs: STRAW-COLORED TO SAFFRON YELLOW, smooth, slender, short. No wax.

Branches: Slender, short, horizontal.

Bole: Straight in alpine races, some crook in eastern European races.

Bark: DARK GRAYISH BROWN to black.

Flowers: *Female: purple, varying* from maroon-black. Male: scarlet or green.

Fruit: LARGE CONES 3/4–1 1/2" (2–4 cm). Scales straight. NOT INCURVED. *Down* on the outside. SCALES ONLY SLIGHTLY OPEN WHEN RIPE.

Remarks: VERY FAST GROWING, LITTLE TAPER. Very tall and slender.

N.T.S. *(G.I.B.)*

Northern White Cedar

Northern White Cedar

(Arbor Vitae)

Thuja occidentalis L.

Range: North of White Mountains.

Habitat: Bogs and swamps and limestone areas.

Profile: *Pyramidal* crown with widely spreading branches in older trees.

Branching: Alternate.

Leading Shoot: Wavy, often drooping.

Leaves: Scale-like, evergreen. Small 1/8" (3–6 mm) long FLAT SCALES CLOSELY OVERLAPPING in 4 rows. Front and back 2-ranked spray flattened with a SINGLE RESIN DOT. Dark *yellow-green.* 1 year seedlings have spreading needles. Persisting on older branches, becoming enlarged and pointed. Very soft foliage.

Buds: Inconspicuous except flower buds.

Twigs: Yellow-brown, flattened, *fan-shaped with resinous taste. Aromatic when crushed.*

Branches: Dark brown.

Bole: *Rapid taper.* Older trees often spiral grained, thin, brown, PEELING OFF IN SHRED-LIKE STRIPS.

Flowers: Olive brown, solitary, on twig tip.

Fruit: Small, OBLONG, DRY, BROWN CONE 1/2" (1–9 mm), *upright on twig.* Few loose scales, falling off after first winter; woody. Seeds with broad WINGS ALL AROUND.

(P.K.)

Atlantic White Cedar

Atlantic White Cedar

(Southern White or Coast White Cedar)

Chamaecyparis thyoides (L.) Bsp

Range: Coastal bogs and swamps. Rare inland.

Habitat: Bogs and river banks.

Profile: *Rounded crown.*

Branching: Alternate.

Leading Shoot: Indistinct.

Leaves: *Scale-like,* overlapping in 4 rows. GLANDULAR ON BACK, *dark blue-green,* turning brown in second year, persistent. Sharp-pointed in seedlings.

Buds: Inconspicuous.

Twigs: Numerous branchlets, forming a flattened spray. Soft, smaller than *Thuja,* aromatic.

Branches: Slender, somewhat drooping.

Bole: *Slight taper.* Thin, BROWN IN BRAIDED RIDGES, SPIRALLY TWISTED.

Flowers: Nearly black, stamens opposite in pairs.

Fruit: GLOBE-LIKE, 1/4" (5–9 mm) with 3 pairs mushroom-like scales. BLUISH-PURPLE SCALES DO NOT OVERLAP. 6 scales, FLESHY-LEATHERY, seated directly on twig. *Splits at maturity.* BLUEBERRY-LIKE when immature.

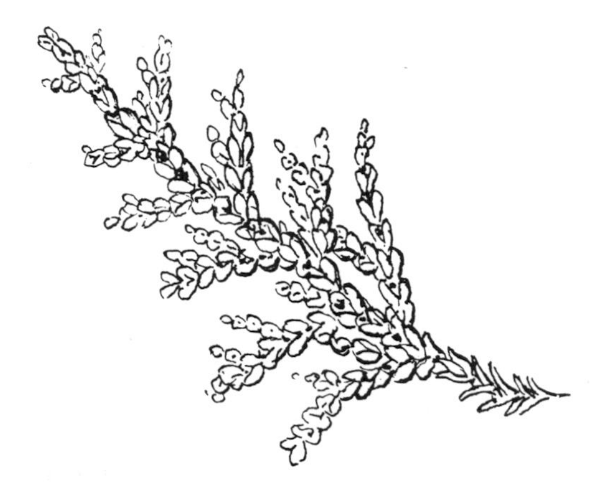

N.T.S. (P.K./G.I.B.)

Atlantic White Cedar (top) & Northern White Cedar (bottom)

Comparison of the White Cedars

Northern White Cedar:

- Yellowish green leaves
- Flattened twigs, fan-shaped
- Oblong cones, thin scales, single resin gland

Atlantic White Cedar:

- Smaller bluish green leaves, glandular
- Twigs only slightly flattened, smaller than above
- Cones spherical on stalks, blue, berry-like
- Cone scales thickened, shield-shaped

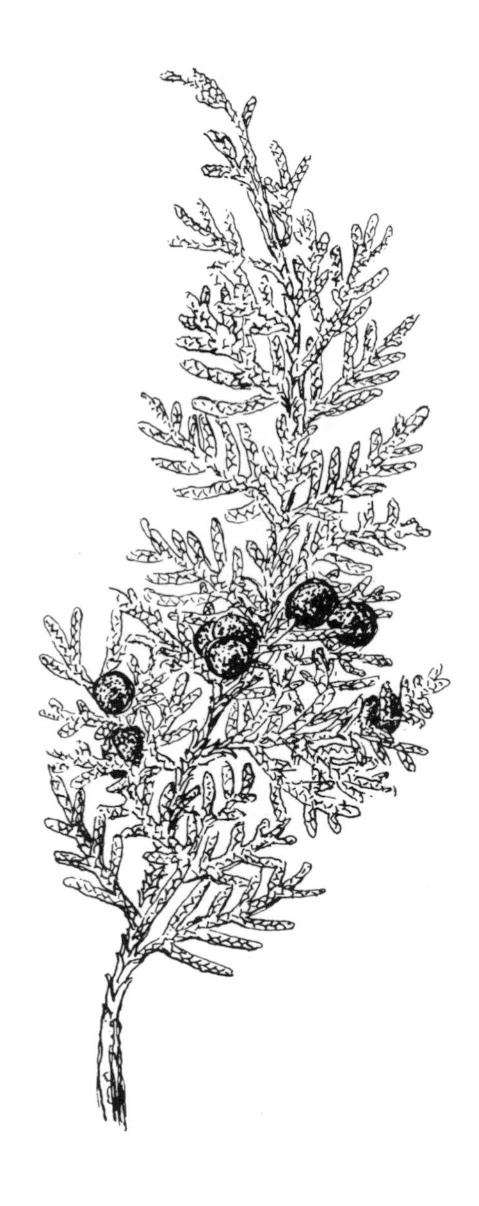

(P.K.)

Northern Red Cedar

Northern Red Cedar

(Savin)

Juniperus virginiana L. var *crebra* Fern. & Grisc.

Range: Eastern U.S. and southern Ontario and New Brunswick.

Habitat: Dry open pastures and waste land.

Profile: Columnar, pointed top, a tree.

Branching: Alternate.

Leading Shoot: ERECT, often drooping.

Leaves: Needles of TWO KINDS. On young branchlets *thin and sharp, like juniper.* On mature branchlets small *scale-like* 1/16" (2 mm) long, *arranged in 4 ranks* on rounded stems.

Buds: Minute, 1/16" (2 mm), inconspicuous.

Twigs: *4-sided in cross section,* WOOD FRAGRANT.

Branches: Spreading, drooping, pendulous.

Bole: *Often fluted at base,* and with double stems.

Bark: Light reddish brown, fibrous, separating *into fringed scales or peeling in long strips.*

Flowers: Minute cones. Male and female on different trees.

Fruit: Berry-like cone, fleshy, bluish bloom, 1/4–1/3" (6–8 mm) diameter, sweet, aromatic.

(P.K.)

Ground Juniper

Ground Juniper

Juniperus communis L. Var *depressa* Pursh.

Habitat: Dry open land.

Profile: Low spreading branches, often ascending. A shrub.

Branching: Alternate.

Leading Shoot: No one leader.

Leaves: All thin, straight AWL-SHAPED, SHARP-POINTED. BROAD WHITE STRIPE ON LOWER SURFACE.

Buds: 1/8" (3 mm) long with pointed scales.

Twigs: Slender, *triangular in section, reddish brown,* with projecting ridges.

Branches: Depressed.

Bole: Many intertwining stems, rarely over 1" (25 mm) in diameter.

Bark: Reddish brown, scaling off in papery sheets.

Flowers: Very small scales. Sexes on different plants.

Fruit: Berry-like, green-gray when young, bluish black *with waxy bloom when ripe.* 1/3–1/2" (6–12 mm), aromatic.

(G.I.B.)

Douglas Fir

Douglas Fir

Pseudotsuga menziesii (Mirb.) Franco

Range: Western U.S. Planted in east.

Habitat: Deep, rich, well-drained, porous loams where there is an abundance of both soil and atmospheric moisture.

Profile: Large tree, pointed crown.

Branching: Opposite.

Leading Shoot: Erect, straight.

Leaves: Rather long 3/4–1 1/4" (2–3 cm), yellowish green or bluish green in some races or varieties, flattened and standing out from all sides of twig. *Slightly stalked,* leaving a round raised scar when falling, *tilted at an angle* to twig. Slightly aromatic.

Buds: Red brown, smooth, sharp, *not resinous.*

Twigs: Reddish or yellowish brown.

Bole: Straight, tapering.

Bark: Smooth on young trees; on old trees divided into thick reddish brown ridges with *deep fissures.*

Flowers: Male: orange-red. Female: slender, red-tinged scales.

Fruit: Oval to cylindrical, pendant with *3-lobed scales or bracts longer than the cone scales.* Matures in 1 year.

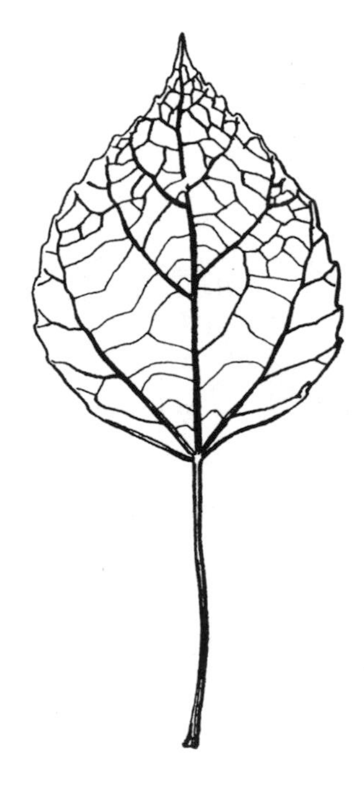

(G.I.B.)

Trembling Aspen

Trembling Aspen

(Quaking Aspen, Popple)

Populus tremuloides Michx.

Range: Northern U.S. and Canada to Alaska.

Habitat: Dry open woods and recent burns.

Profile: Tall, straight, slender, small crown, medium-sized tree.

Branching: Alternate.

Leaves: Stem slender and *flattened* leaves *circular,* rounded with a short abrupt point. As long or longer than broad. VERY FINELY AND EVENLY TOOTHED. 2–8 cm long; 1.8–7 cm broad.

Buds: Slender, reddish brown shining 1/4" (.6 cm), *smooth. Terminal bud present.*

Twigs: Smooth, reddish brown on short shoots. Greenish stripes or dots. *Pith star-shaped.*

Bole: Straight, little taper.

Bark: Smooth, green with dark patches. Whitish bloom on young trees, becoming rough on older trees.

Flowers: Male and female on separate trees. Drooping catkins, 3–5" long bearded segments.

Fruit: Capsule 1/4" (.6 cm), long small seeds with silky hairs.

(G.I.B.)

Large-Toothed Aspen

Large-Toothed Aspen

(Big-Toothed Aspen)

Populus grandidentata Michx.

Range: Southern Canada, Lake States to southern mountains of New England.

Habitat: Better soils and burns.

Profile: Larger crowned, taller than trembling aspen, large tree up to 2' diameter.

Branching: Alternate.

Leaves: Circular, *larger.* (4–12 cm) WITH COARSE WIDELY UNEQUAL SEPARATED TEETH. (5–15) *Young leaves covered with white woolly hairs or felt.* Stem flattened as long as diameter of leaf.

Buds: Dull, grayish, *coated with thin white felt.*

Twigs: Brownish gray with star-shaped pith.

Bole: Straight, little taper.

Bark: Smooth, olive green, later becoming DEEPLY FURROWED. *More yellow* than trembling aspen.

Flowers: Male and female on separate trees. Flowers coarser than trembling aspen, somewhat hairy or with bloom.

Fruit: Capsule slightly hairy or with bloom. Similar to Trembling Aspen.

(G.I.B.)

Cottonwood

Cottonwood

(Carolina Poplar)

Populus deltoides Marsh.

Range: Along Connecticut River and southward.

Profile: Crown oblong or narrowly oval, large tree, largest of the poplars and aspens.

Branching: Alternate.

Leaves: Large *triangular* with flattened stalks and coarse *rounded teeth* 2–4" (6–12 cm) long. Stem flattened at base of leaf. Leaves from sprouts very large, *smooth* lustrous green.

Buds: Brown, shiny, 1/2–1" (1.5–2.5 cm) long, 6–7 scales, heavy, sticky with glue, fragrant, yellow resin inside *curved.*

Twigs: Smooth, *yellow* brown.

Bole: Straight, tapering, horizontal branches; few lower branches.

Bark: Greenish yellow on young tree; gray, furrowed on older trees.

Flowers: Drooping, loosely grouped catkins April–May.

Fruit: Oval capsule, 1/3" long; 3–4 valved smooth borne like a string of beads.

(G.I.B.)

Lombardy Poplar

Lombardy Poplar

(Poplar-Pine)

Populus nigra L. *var italica* Du Roi Muench.

Range: Introduced from Europe.

Habitat: Propagated by cuttings only.

Profile: Oblong to tall green column, SPIRE-LIKE, BRANCHES ERECT AND CLOSE TO MAIN TRUNK.

Branching: Alternate.

Leaves: Small, *triangular,* as broad as long. *Coarse teeth.* Somewhat 4-sided, sides of base slope upward. 2–4" (5–10 cm).

Buds: Reddish, out-curving, sticky. Smaller than Carolina Poplar.

Twigs: Smooth, orange to ashy gray. Slender, pressed to trunk.

Bole: Straight, tapering covered by branches to the ground.

Bark: Deeply *furrowed,* gray to brown

Flowers: Infertile.

(G.I.B.)

Balsam Poplar

Balsam Poplar

Populus balsamifera L.

Range: Northern New England and Canada, west to Lake States, Rockies, Alaska.

Habitat: Riverbanks and gravels,

Profile: Medium to large tree, narrow open crown.

Branching: Alternate.

Leaves: 3–6" long (7–15 cm), tapering to a *sharp point.* Smooth above, pale beneath with a metallic luster. Finely toothed, with *minute hairs on teeth.* Rounded at base. Leaf stalks ROUNDED.

Buds: Long, sharp-pointed 3/4" (15–25 mm) STICKY WITH YELLOW GUM WITH BALSAM FRAGRANCE. Terminal bud with 5 scales.

Twigs: Bright reddish brown, smooth, shiny, becoming dark orange-gray tinged with yellow green on older twigs.

Bole: Cylindrical, straight.

Bark: Light brown tinged with red; on older trees dark gray in rounded ridges. *Bark smooth on upper part* of tree.

Flowers: Catkins fringed at broad summit by bristle-like spines.

Fruit: Large drooping catkins, ripening in early spring. Seeds ovoid.

(G.I.B.)

Willow

Willow

Salix Sp.

Range: Northeast U.S. and Canada.

Habitat: Damp ground.

Profile: Wide-spreading crown, bole crooked or leaning.

Branching: Alternate.

Leaves: Long, thin, narrow in most species, usually smooth, often silvery beneath.

Buds: Alternate with a SINGLE CAP-LIKE SCALE. NO TERMINAL BUD.

Twigs: Slender, often pale green, red or orange. *Drooping.*

Bark: Smooth when young, often yellow, becoming furrowed in older trees.

Flowers: Male and female catkins on separate trees. Pussy willows are flower buds.

Fruit: Small capsule discharging silky-haired seeds that float in air.

Remarks: There are many species of willow that hybridize freely. See other manuals for species. Black willow: *Salix nigra.*

(G.I.B.)

Sweet Gale

Sweet Gale

Myrica gale L.

Range: Southern New England and Mid-Atlantic States.

Habitat: Shallow water and swamps and shores.

Profile: *Spreading bush* 1–4' (2 m) high.

Branching: Alternate, twisted, turning upward.

Leading Shoot: Erect.

Leaves: 1–2" (2–5 cm) LONG, NARROW, POINTED, grayish. Fragrant. BROADEST NEAR TIP WITH FEW TEETH AT TIP. Hairy on both sides with resin dots. *Wedge-shaped base.*

Buds: Pointed. Flower buds with MANY OVERLAPPING SCALES. No terminal bud.

Twigs: Purple-brown, smooth or rusty WITH SMALL YELLOW SPOTS.

Bark: Brown.

Flowers: Male flowers short, brownish, scaly catkins. Female, conelike, appearing in April, remaining all summer. Fragrant when crushed.

Fruit: Small persistent cones 1/2" (10–12 mm) *cone-like bunches at the end of branchlets.*

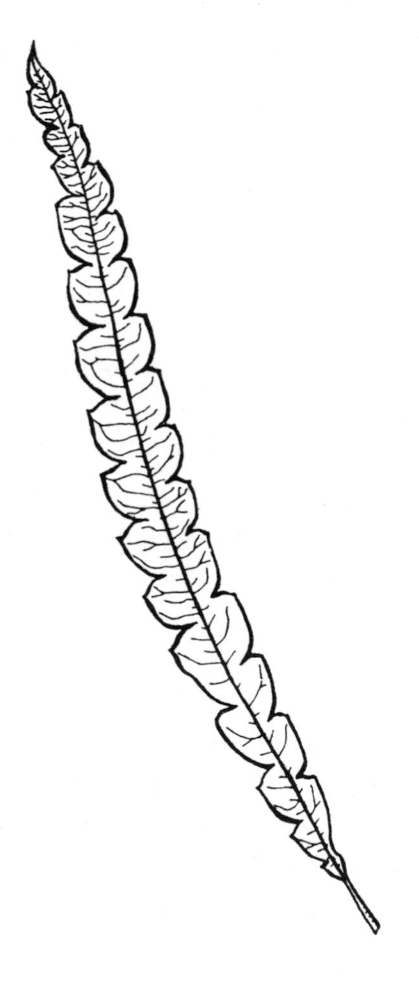

(G.I.B.)

Sweet Fern

Sweet Fern

Comptonia perigrina (L.)

Range: Nova Scotia westard to Saskatchewan, and southward to North Carolina, Indiana and Minnesota.

Habitat: Open dry sandy roadsides and pastures.

Profile: Low, spreading, often depressed 1–3' (1/2 m) *mat-forming.*

Branching: Alternate, much branched.

Leading Shoot: Drooping.

Leaves: FERN-LIKE, dark green above, pale beneath. DEEPLY LOBED, rounded, hairy with resin dots. 3–6" (7–15 cm) long. Fragrant. 1 cm wide.

Buds: Hairy, few (about 5) *visible scales* very small, rounded. No terminal bud.

Twigs: Gray, with rusty wool when young. Slender, slightly zigzag.

Bark: Brown.

Flowers: Male and female usually on separate plants, appearing before the leaves. Brownish catkins prominent in winter.

Fruit: *Cylindrical,* small shiny nut within burr-like scales.

N.T.S. (G.I.B.)

Bayberry or Wax Myrtle

Bayberry or Wax Myrtle

Myrica pensylvanica Loisel.

Range: Eastern Canada south to North Carolina.

Habitat: Wet, poor soil near seacoast.

Profile: Medium sized shrub, 1–6' (.3–2 m).

Branching: Alternate.

Leaves: Blunt-pointed, oval, aromatic smell. Narrow, 1–3" (2–7 cm) long, thin, few or no teeth, on short stalks pale beneath with scattered wax dots. Hairy along the veins.

Buds: Globular, pale, reddish brown, smooth with long black hairs dotted with yellow resin glands.

Twigs: Stiff, whitish gray or grayish brown. Thick ends of twigs with yellowish resin globules. Often hairy.

Bark: With short longitudinal cracks.

Flowers: Male and female flowers on separate plants, male catkins enclosed in winter buds, inconspicuous. Female flowers in crowded clusters below leafy tips becoming hard white berries.

Fruit: Bony round nuts covered with white or gray wax. Young nuts are densely hairy. Fruit on female plants only-

Butternut

Butternut

(White Walnut)

Juglans cinerea L.

Range: Southern Canada to southern and middle western U. S.

Habitat: Rich woods.

Profile: Large tree.

Branching: Alternate.

Leading Shoot: Straight.

Leaves: Compound with *7–17* LEAFLETS; leaf 1½' (50 cm), hairy beneath leaf stalks and *covered with sticky hairs*. Prominent leaf scars with *hairy fringe* on upper margin. Leaflets oblong.

Buds: Soft, hairy, angled, not round, gray-brown. Terminal buds large, longer than broad, blunt-pointed.

Twigs: Stout, greenish gray to reddish brown with streaks (lenticels), white spots, pith chambered with *thick* diaphragms, *dark chocolate brown to black.*

Branches: Thick, horizontal.

Bole: Straight, clear of branches.

Bark: Light gray, smooth when young breaking into flat braided ridges.

Flowers: Catkins, brown, drooping, 8–12" (20–30 cm) stamens.

Fruit: Elliptical, sticky, pointed 2" (5 cm), green nut *deeply* corrugated with *sharp ridges*. Seed sweet, *very oily.*

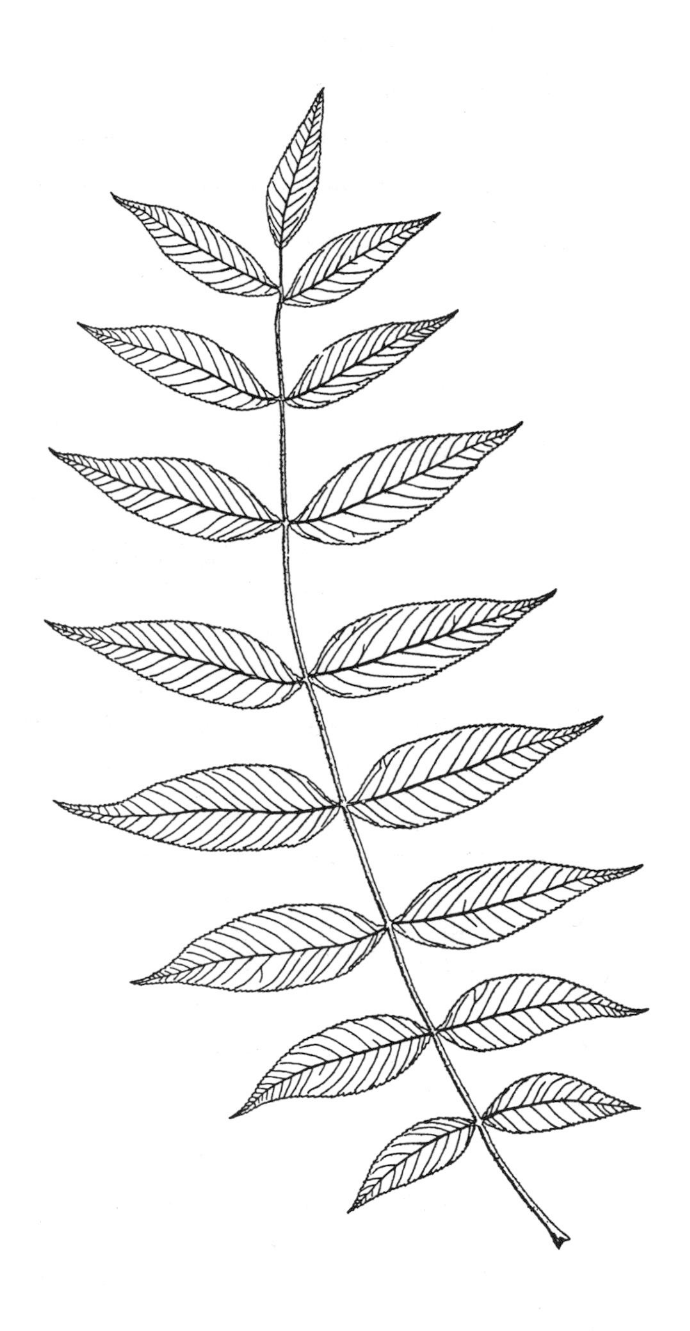

N.T.S. (G.I.B.)

Black Walnut

Black Walnut

Juglans nigra L.

Range: Western Massachusetts to Minnesota, south to Florida.

Habitat: Rich woods

Profile: Very large tree.

Branching: Alternate.

Leading Shoot: Straight.

Leaves: Compound with *11–17 leaflets* (always more than butternut), slightly downy beneath 2' long, terminal leaflets suppressed. Leaf scars notched, *no hairy fringe.* Leaflets oval.

Buds: Grayish, pale, silky to downy, not longer than broad, flattened, lateral buds opening slightly during winter.

Twigs: Light brown to orange brown pith chambered with *thin* diaphragms, *buff colored.* Acrid taste.

Branches: Thick, horizontal.

Bole: Straight, clear of branches

Bark: Dark brown to grayish black, divided into deep, narrow furrows with thin ridges, forming a diamond-shaped pattern.

Flowers: Catkins with 20–30 stamens.

Fruit: Round or nearly so, 2–2 1/2", nut corrugated, with rounded ridges. Seed sweet, oily.

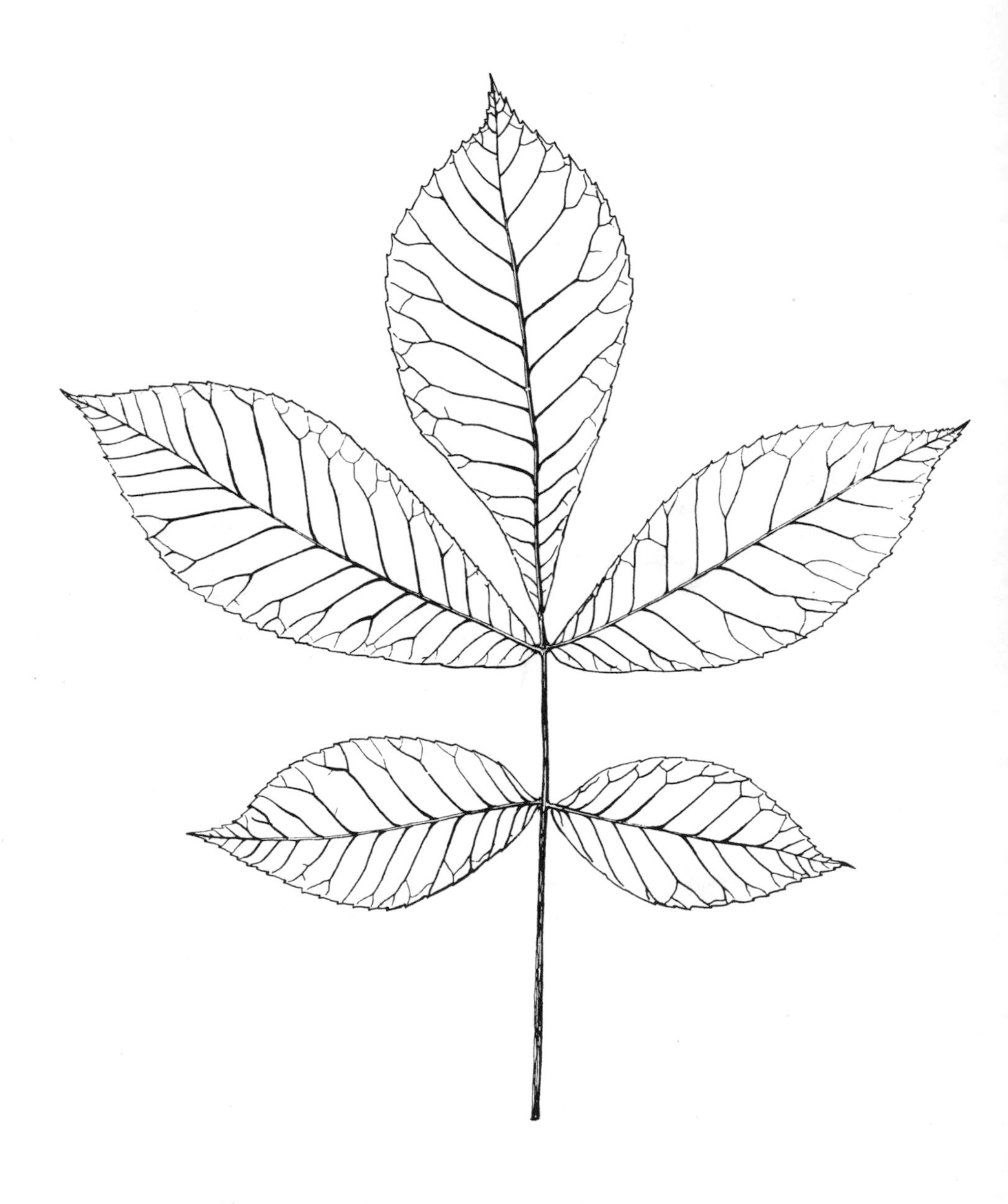

N.T.S. (G.I.B.)

Shagbark Hickory

Shagbark Hickory

(Shellbark Hickory)

Carya ovata (Mill.) K. Koch

Range: Southern New England. Southern Maine, Quebec, Ontario, west to central states, south to Florida and Texas.

Habitat: Rich woods.

Profile: Large tree.

Branching: Alternate.

Leading Shoot: Straight.

Leaves: 4–6" (10–15 cm) compound with 5 (sometimes 7) leaflets, with the 3 upper leaflets *much larger and longer than the lower terminal leaflet.* Slightly toothed, with a *tuft of hairs on each* tooth.

Buds: Dark brown, long-pointed 1/2" *or more* (1 cm) thick, (1.3–2.5 cm) long. Scales *remaining over winter.*

Twigs: Very tough and hard to break. Very *stout,* pith solid .

Branches: Ascending, forming a dense head.

Bole: Often free of branches for 50 feet.

Bark: Smooth when young, splitting into long plates that curve outward, free at both ends.

Flowers: Male and female on same tree. Male flowers 3-branched catkins; female in short spikes.

Fruit: *Thick* globe-like husks *split away* when ripe. Nut with *4 angles,* whitish brown shell, thick, kernel sweet.

(P.K.)

Pignut Hickory

Pignut Hickory

(Sweet Pignut)

Carya glabra (Mill.) Sweet

Range: Southern New England southward.

Habitat: Dry woods and slopes, warm and rocky slopes.

Profile: Medium sized tree.

Branching: Alternate.

Leading Shoot: Straight

Leaves: Compound with 5 leaflets as in shagbark, but *smaller* and *without* minute tufts *of hair on teeth.*

Buds: *Less than* 1/2" *long* (9–12 mm), pale brown outer scales, dark bud scales dropping in autumn, leaving silky-hairy covering.

Twigs: As tough as shagbark. *Slender* smooth, reddish brown to gray, pith solid.

Branches: Slender, more or less contorted lower branches, especially droop;ng.

Bole: Shorter than shagbark.

Bark: Less flaky, with rounded ridges.

Flowers: Male and female on same tree similar to shagbark.

Fruit: *Thin* husks *adhere to nut* when ripe. Nut shell *scarcely ridged,* kernel sometimes bitter.

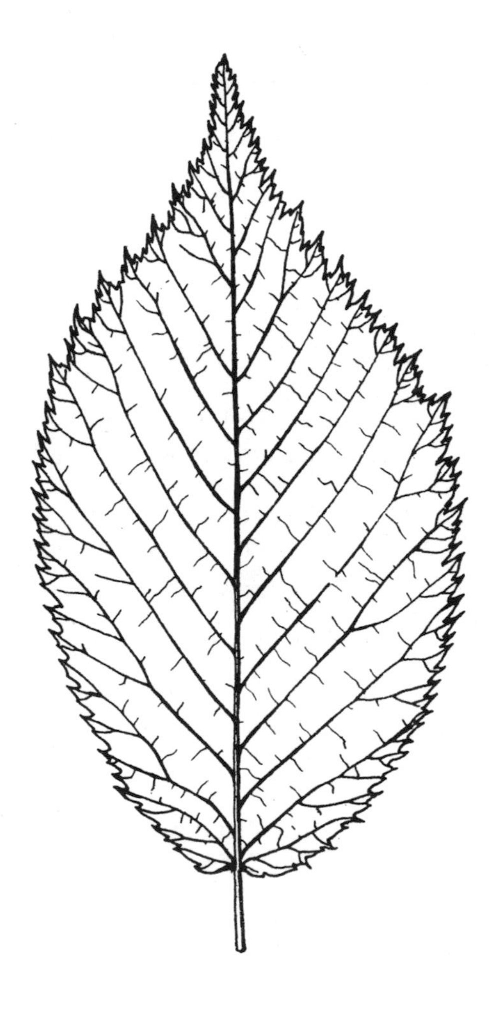

(G.I.B.)

Hop Hornbeam

Hop Hornbeam

(Leverwood, Ironwood)

Ostrya virginiana (Mill.) K. Koch.

Range: Nova Scotia westward to southern Manitoba, Northern Minnesota, the Black Hills of Dakota, eastern Nebraska and Kansas, and south to Florida and west to eastern Texas and Oklahoma.

Habitat: Dry upland hillsides and pastures.

Profile: Round, spreading crown

Branching: Alternate.

Leading Shoot: Inconspicuous.

Leaves: Sharp-pointed, doubly toothed, 1–4" (2.5–10 cm), slightly *downy on both sides.*

Buds: Ovate, 2–4 mm, reddish brown, sharp, *tinged with green,* terminal bud lacking.

Twigs: Slender, *jointed in zigzag fashion,* smooth or with few hairs

Branches: *Zigzag.*

Bole: Straight, small taper.

Bark: *Scaly,* thin, *gray with long flaky scales, upturning on edges, clinging in the middle.*

Flowers: Male catkins 2" (.5 cm), appear with leaves; female small.

Fruit: HOP-LIKE CLUSTER OF BLADDERY sacs.

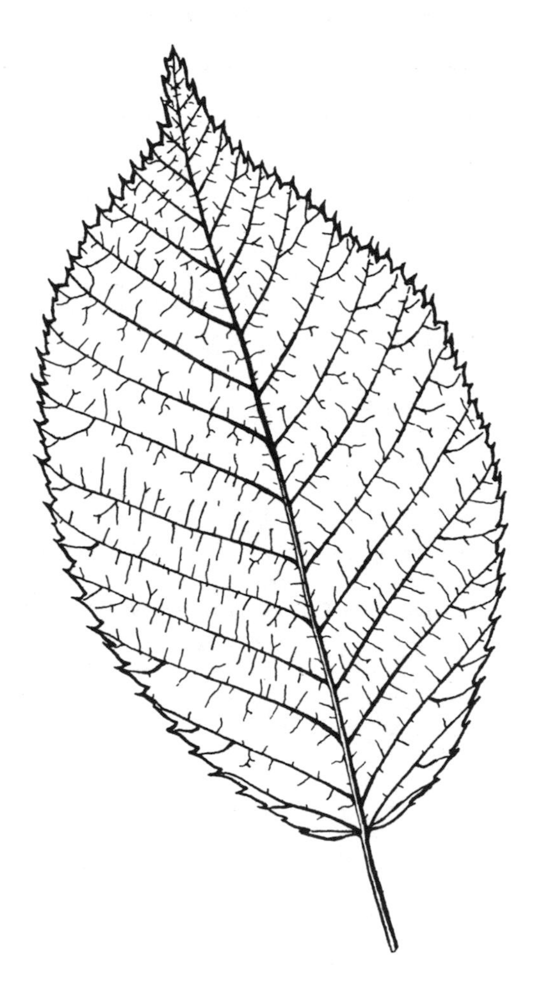

(G.I.B.)

American Hornbeam

American Hornbeam

(Ironwood, Blue Beech)

Carpinus caroliniana Walt.

Range: Nova Scotia south to Florida, westward to southern Ontario, northen Minnesota, eastern Nebraska, Kansas, Oklahoma and Texas.

Habitat: Rich damp woods, stream bottoms.

Profile: Round spreading crown often flat-topped.

Branching: Alternate.

Leading Shoot: Inconspicuous.

Leaves: Oblong, sharply and finely toothed, 1–4" (2.5–10 cm), *entirely smooth above,* slightly downy below.

Buds: Ovate, 2–4 mm, *reddish brown, angled,* scaly.

Twigs: Slender (1 mm toward end) *dark red,* shining smooth or slightly hairy.

Branches: *Wavy.*

Bole: Often crooked, tapering.

Bark: *Smooth, gray-green, furrowed, fluted or "muscled," wavy.*

Flowers: Male catkins 1" (2.5 cm); female small, appearing with leaves.

Fruit: SMALL NUT enclosed in life-like, 3-LOBED pod. Catkins absent in winter.

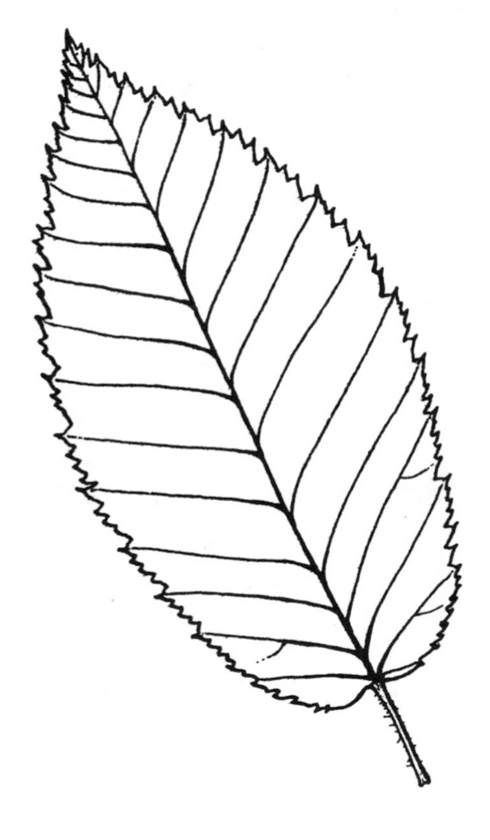

(G.I.B.)

Yellow Birch

Yellow Birch

(Silver Birch)

Betula lutea Michx f.

Range: Southern Canada, Lake States, northeast and Appalachian Mountains.

Habitat: Sandy loam soils in a mixture of sugar maple and beech.

Profile: Large, tall tree of the north woods.

Branching: Alternate

Leading Shoot: Straight

Leaves: *Long, oval,* sharply double-toothed. Dull, dark green above, lighter below, with slight hairs in veins, base rounded to near heart-shape.

Buds: Sharp, reddish brown, appressed, terminal bud absent, *ovate,* with chestnut-brown scales.

Twigs: Stout, bronze or *bright brown, shiny,* lenticel dots indistinct. *Wintergreen odor and taste.*

Bark: *Yellowish or silver gray, peeling in thin strips* when young. Rough and platy on old trees. Ribbon-like strips of young bark roll back in curls.

Bole: Straight, long tapering, cylindrical.

Flowers: Male and female in separate catkins on the same tree. Male catkins present in winter.

Fruit: Catkins rounded, *globe-like,* scales somewhat *hairy,* erect on twig, 3/4" long (2 cm), catkins persist in winter.

Remarks: Most like red birch.

(G.I.B.)

Black Birch

Black Birch

(Sweet Birch, Cherry Birch)

Betula lenta L.

Range: Appalachian Mountains and adjacent regions; also Michigan.

Habitat: Deep, rich, moist but well-drained soils. Also found on rocky sites.

Profile: Tall, slender tree.

Branching: Alternate

Leading Shoot: Straight

Leaves: *Heart-shaped,* sharply double- or single-toothed, smooth above with tufts of white hair on the veins beneath, dull, dark green appear opposite on shoots. *Aromatic.*

Buds: Sharp, reddish brown, *long* terminal bud absent on long shoots. Bud scales downy on margins.

Twigs: Smooth, slender, light reddish brown with very fine horizontal lines. VERY STRONG WINTERGREEN ODOR AND TASTE.

Bark: *Dark brown,* smooth, close, *not peeling.* Long horizontal stripes (lenticels) resembling cherry. Older trees with thick plates. Very dark. *Inner bark with wintergreen taste.*

Bole: Straight, small taper.

Flowers: As in Yellow Birch.

Fruit: Catkins *cylindrical,* short, scales smooth about 1" long (2.5–3 cm).

Remarks: Resembles black cherry.

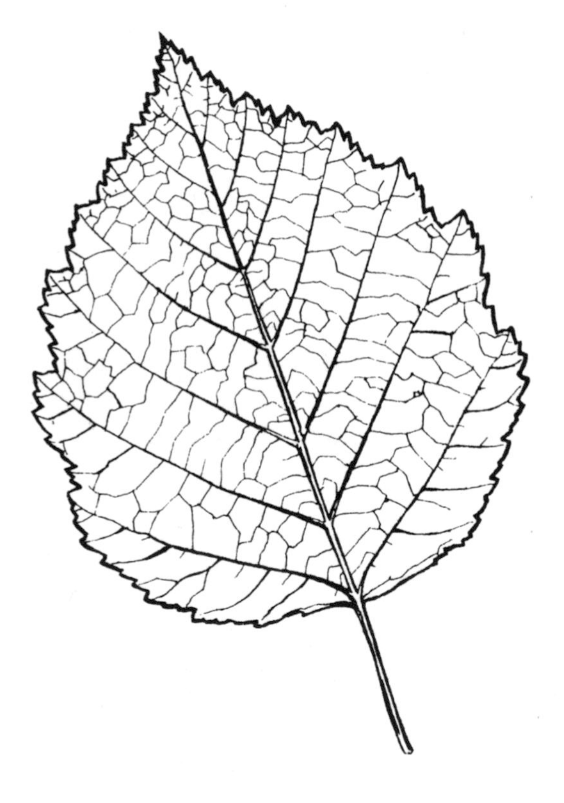

(G.I.B.)

Paper Birch

Paper Birch

(Canoe Birch)

Betula papyrifera

Range: Northeast U.S. and Canada.

Habitat: Moist, mineral soils. Quickly establishes cover in cut-over and burned lands.

Profile: Large, tall, straight tree.

Branching: Alternate.

Leading Shoot: Straight.

Leaves: *Broad,* rounded at base, often heart-shaped, TIPS BLUNT. Irregularly *double-toothed* on margins. Alternate, 2–3" (5–7 cm) long. Somewhat STICKY beneath.

Buds: Chestnut brown, with scales, *downy on margins.*

Twigs: Fairly *stout, more or less hairy,* dark reddish brown, with *white dots or horizontal oblong patches.*

Bole: Erect, straight, tall, little taper.

Bark: *Reddish when young,* white peeling, *outer layers easily separating* into thin sheets. Delicate pinkish or yellowish tinge when not exposed to sun. Conspicuous horizontal streaks.

Flowers: Brownish catkins in twos or threes. Male and female in separate catkins on the same tree. Male 3" long.

Fruit: Short-stalked cylindrical catkin, 1–2" (2–5 cm) long.

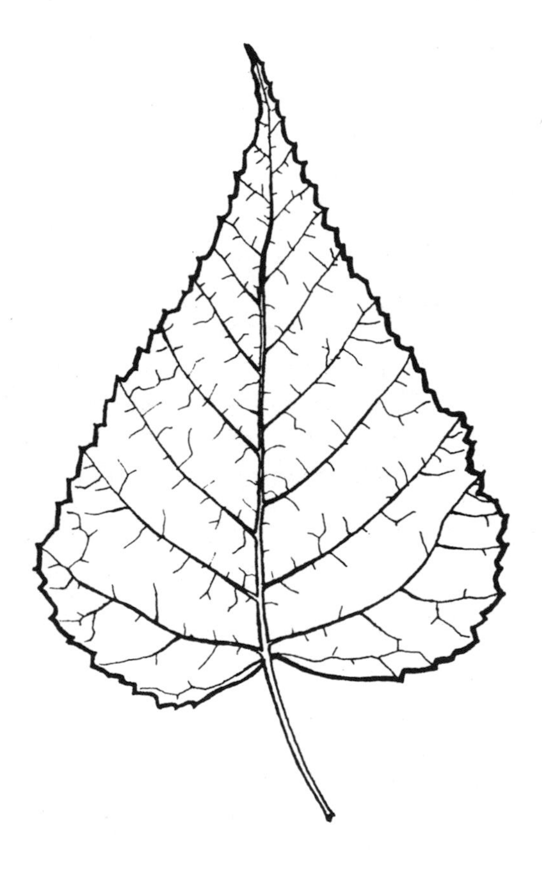

(G.I.B.)

Gray Birch

Gray Birch

(Old Field Birch, White Birch)

Betula populifolia

Range: Northeast U. S. and South Canada.

Habitat: Abandoned farms and burned-over lands; it will grow on the poorest of sterile soils.

Profile: Small, low, often in clumps, bending.

Branching: Alternate.

Leading Shoot: Bending.

Leaves: SMALL, TRIANGULAR, very *long tapering tip, finely* double-toothed. Shiny smooth above, tufts of hairs on veins below. Appear to be *opposite or whorled.* Unpleasant taste.

Buds: Very small, short, light brown to greenish terminal bud lacking except on spur shoots.

Twigs: *Fine,* reddish to yellow brown, SMOOTH, roughened with white dots, dense, close, bushy. Outer shoots *very thin, wiry.* Inner twigs warty.

Bole: Bending, often bent over.

Bark: *Dark brown when young.* Close, NOT PEELING, *chalky* white with *black bark blotches.*

Flowers: Brownish, usually solitary new male catkins present in winter, developing with the leaves in May.

(G.I.B.)

European White Birch

European White Birch

Betula alba

Range: Europe.

Habitat: Dry soils. Introduced from Europe.

Profile: Planted as a shade tree, U.S.

Branching: Alternate.

Leading Shoot: Often drooping.

Leaves: Oval, sharp pointed. Leaves tremulous like the Aspens.

Twigs: Drooping. Resin glands, smooth, variable.

Bole: Erect, tapered.

Bark: Creamy white, often dirty black, deep furrowed when old with gusset-like folds.

(G.I.B.)

River Birch

River Birch

(Red Birch)

Betula nigra L.

Range: Rare in New Hampshire, common in Atlantic states and westward.

Habitat: Lake and stream shores.

Profile: Slender tree, usually short and much branched.

Branching: Alternate.

Leading Shoot: Curving.

Leaves: Diamond-shaped, wedge-shaped at base, irregularly toothed, pale beneath. Somewhat lobed, some deeply and doubly toothed. Dark green above, bluish or downy beneath.

Twigs: Slender, dark red with white spots. No wintergreen odor or taste, often downy or hairy. Lenticels prominent.

Bole: Short, much branched.

Bark: *Peels freely.* Brown to red, or greenish brown. Peels in very thin layers salmon-pink. Curls back and remains several years as ragged fringes. Inner layers light pinkish.

Flowers: Resembles black birch; appear in April.

Fruit: Only birch maturing fruit in spring. Cylindrical, erect catkin, hairy, scales falling.

Remarks: Only birch found at low elevations in the south.

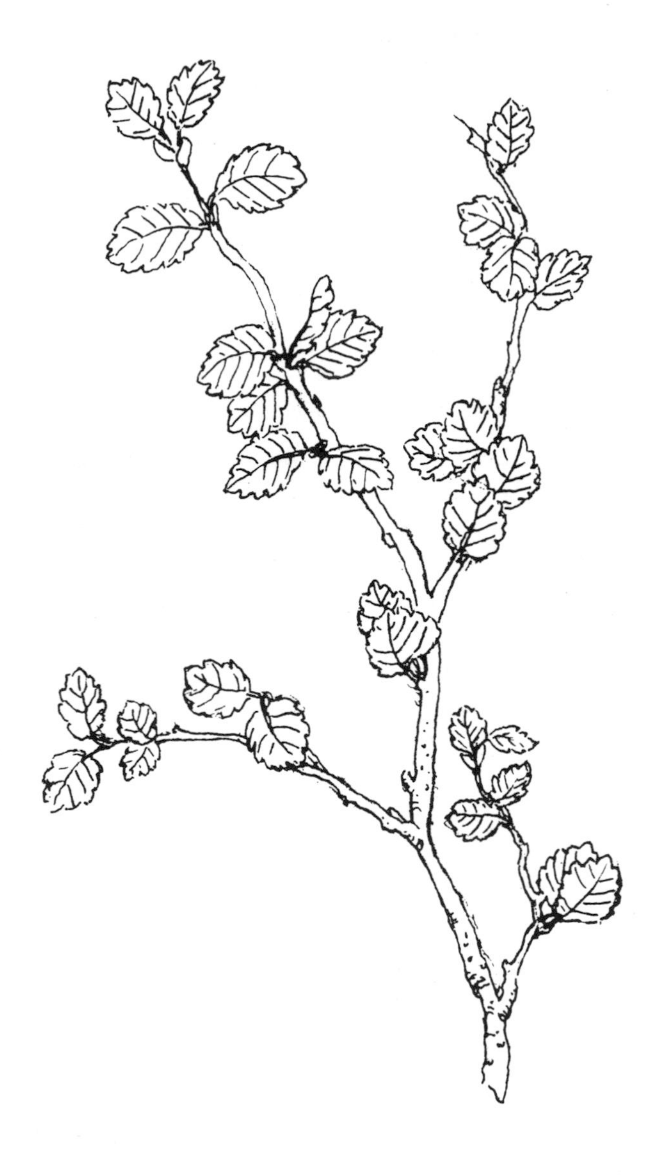

(P.K.)

Dwarf Birch

Dwarf Birch

(Alpine Birch)

Betula minor (Tuckerm.) Fern.

Range: New England and New York.

Habitat: Alpine zone of high mountains.

Profile: Low shrub.

Branching: Alternate.

Leading Shoot: Drooping to slanting.

Leaves: Very small, 1–2" long (2.5–5 cm) rounded or wedge-shaped at base. *Sharp-pointed,* resembling Gray Birch.

Buds: Pointed or blunt, sticky.

Twigs: Close, dark, tough, smooth, often bearing glandular warts.

Bole: Bushy.

Bark: Dark.

Flowers: Flower in June. Catkins 3/4–2" (2–5 cm) long.

(G.I.B.)

Speckled Alder

Speckled Alder

Alnus rugosa (DuRoi) Spreng.

Range: Common in northern New England.

Habitat: Swamps, margins of streams, lakes.

Profile: Spreading large shrub in clumps up to 18' high (5 m).

Branching: Alternate.

Leaves: Oval or rounded 2–4" (5–10 cm) long. Broadest below middle. *Doubly toothed.* Green above. PALE BENEATH. *Leaf base rounded or heart-shaped, not sticky, cross veins prominent.*

Buds: *Buds stalked,* covered by 2–3 scales equal in length, coated with grayish or brownish down, no terminal bud.

Twigs: Olive-brown to reddish brown, downy. Pith triangular, greenish.

Bark: Brown, with *whitish horizontal dots* (lenticels) *7 mm or more long.*

Flowers: Developed long before the leaves. Male catkins: long, drooping. Female: *recurved,* arching, appearing below male. *Both droop.*

Fruit: Female catkins ripening into 1/2" (1 cm) cylindrical cones persistent in winter.

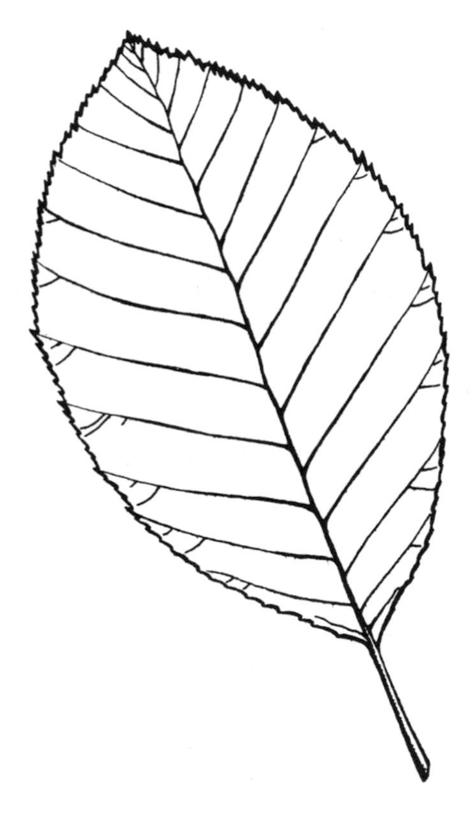

(G.I.B.)

Smooth Alder

Smooth Alder

(Common Alder)

Alnus serrulata (Ait.) Willd.

Range: Southern type north to central New Hampshire.

Habitat: Swamps, margins of streams and lakes.

Profile: Spreading large shrub in clumps up to 18' high (5 m). Shrub or tree-like.

Branching: Alternate.

Leaves: Broadest above middle, single-toothed fine teeth GREEN BOTH SIDES, *wedge-shaped at base, sticky, aromatic* when young, *cross veins weak.*

Buds: Buds stalked, covered by 2–3 scales equal in length, coated with grayish or brownish down, no terminal bud.

Twigs: Slender, greenish-reddish brown, less downy. Pith triangular, greenish.

Bark: *Faint, smaller white lines or none.*

Flowers: Developed long before the leaves. Male catkins: longer than Speckled Alder, drooping. Female: recurved, arching, *erect* above male.

Fruit: Female catkins ripening into 1/2" (1 cm) cylindrical cones persistent in winter.

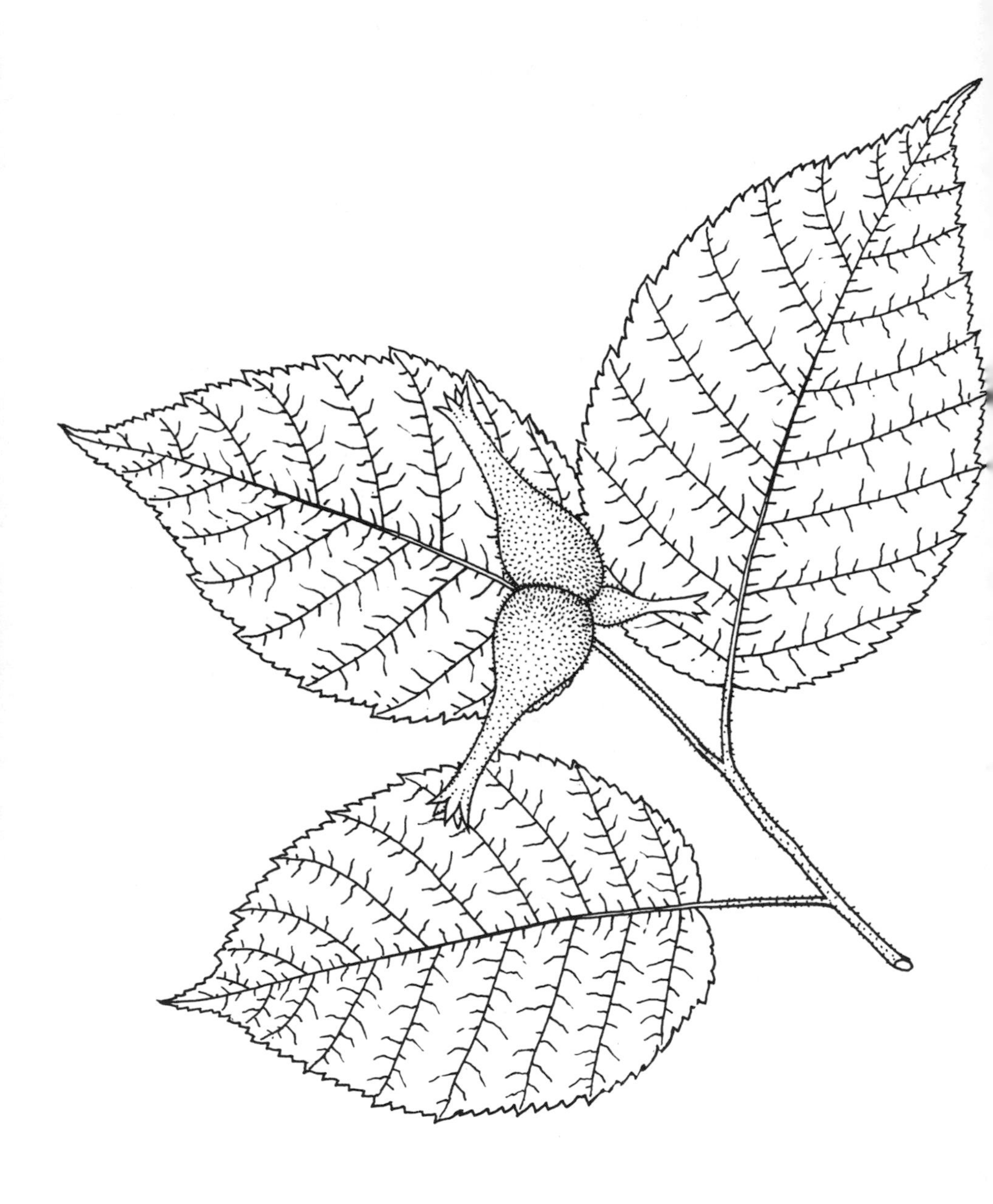

N.T.S. (G.I.B.)

Beaked Hazelnut

Beaked Hazelnut

Corylus cornuta Marsh.

Range: New England.

Habitat: Dry places.

Profile: Small shrub, 3' (1 m).

Branching: Alternate.

Leaves: Irregularly toothed or double-toothed, 2–5" (5–12 cm) long, heart-shaped at base, downy beneath, pale green, upper surface roughened.

Buds: Blunt, pale, downy at tip with 4 exposed bud scales. Terminal bud absent.

Twigs: *Smooth, no hairs or glands.* Slender, more or less zigzag, dull brown.

Flowers: Male catkins shorter than American hazelnut.

Fruit: Nut enclosed in a long, bristly tube-like beak, twice the length of the nut. Beak 4–7 cm.

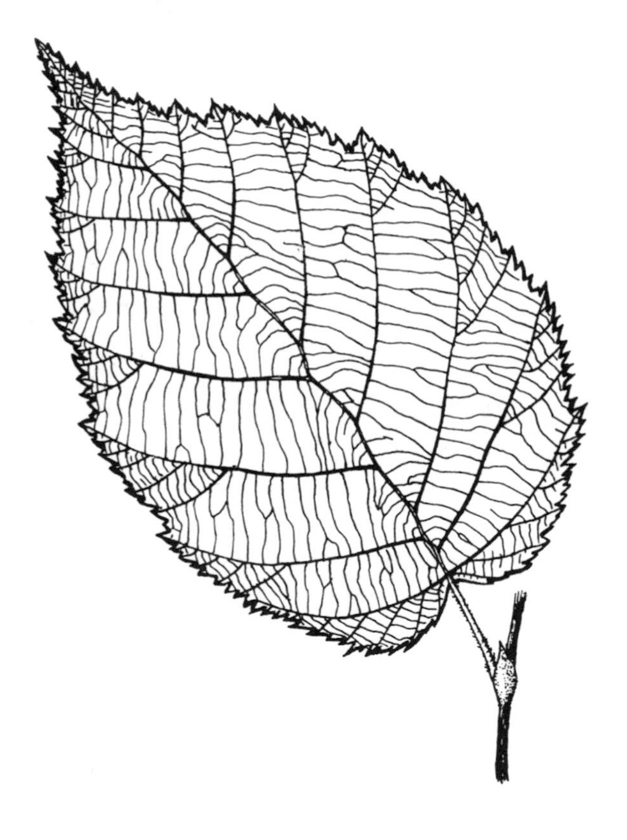

(G.I.B.)

Hazelnut

Hazelnut

(American Hazel)

Corylus americana Walt.

Range: Eastern U.S. and central states.

Habitat: Thickets, open scrub.

Profile: Medium shrub, 3' (1 m).

Branching: Alternate.

Leaves: Curving. Oval, pointed, *double-toothed,* somewhat downy beneath. Heart-shaped at base. Lower surface and stems somewhat hairy.

Buds: *Rounded at tip,* purple brown, somewhat hairy. No terminal bud.

Twigs: Slightly hairy. Hairs on twigs, red when young. *Glands.* Slender, somewhat zigzag, brownish gray.

Flowers: Smooth, thin, dark brown.

Fruit: Male catkins: long, drooping, 3". Female: very short with crimson, hair-like projections. Edible, not surrounded by leafy part. Ripening in July, flared and lobed at tips. 3–5" in a cluster. Hard shelled nut.

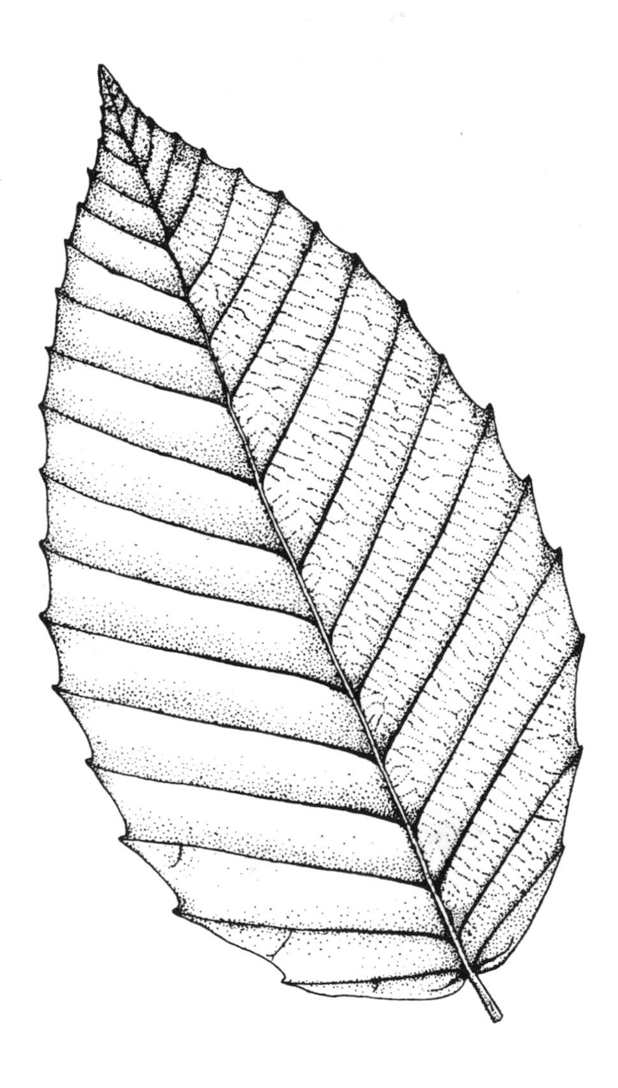

(G.I.B.)

Beech

Beech

(American Beech)

Fagus grandifolia Ehrh.

Range: South Canada to Virginia, west to Lake States.

Habitat: Rich uplands, often dry,

Profile: Large tree with broad crown.

Branching: Alternate.

Leaves: Toothed with *parallel veins terminating in teeth.* Stems hairy. Shiny, papery tough, smooth. *Dry leaves remain on tree in winter.* Yellow in autumn.

Buds: Long, slender, sharp-pointed, 3/4" (2–3 cm), chestnut brown with *overlapping scales.*

Twigs: Alternate, slender, greenish brown.

Bole: Straight, short, much branched at top.

Bark: Smooth, light blue-gray, GRANULAR, *does not change as tree grows older.*

Flowers: Slender catkins in a rounded head, both sexes on the same tree,

Fruit: *Small,* prickly burr, *only 3/4" (1 cm)* in diameter, holding 2 *triangular* nuts.

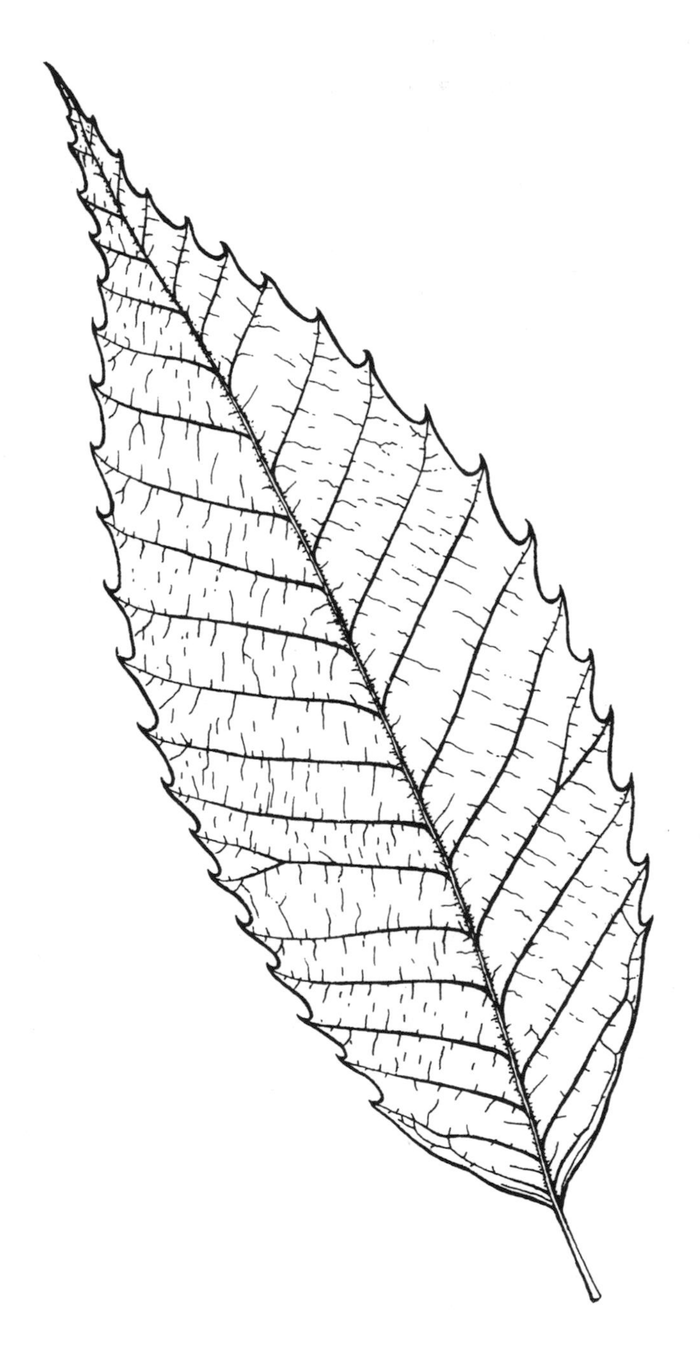

(G.I.B.)

Chestnut

Chestnut

(American Chestnut)

Castanea dentata (Marsh.) Borkh.

Range: South Canada to Georgia, Florida, west to Lake States.

Habitat: Dry, gravel or rocky soil. Mostly destroyed by the bark disease. Persisting as sprouts, soon killed.

Profile: Formerly a very straight, slender, large tree.

Branching: Alternate.

Leaves: *Longer and narrower* than beech, 6–10" (15–25 cm) long, 2" (5 cm) wide. *More deeply toothed,* coarse, pointed teeth, tapered to a point at base. *Smooth and green on both sides.*

Buds: Very short. Terminal bud lacking.

Twigs: Light brown with fine white dots, short twigs go out at right angles.

Bole: Slender, straight, little taper.

Bark: Smooth, dark brown with broad flat ridges, *becoming rough.*

Flowers: Cream-colored catkins, drooping. Both sexes on the same tree.

Fruit: *Large* prickly burr 1–3" (2–7 cm), 2–3 or more nuts, *flattened* on one or both sides.

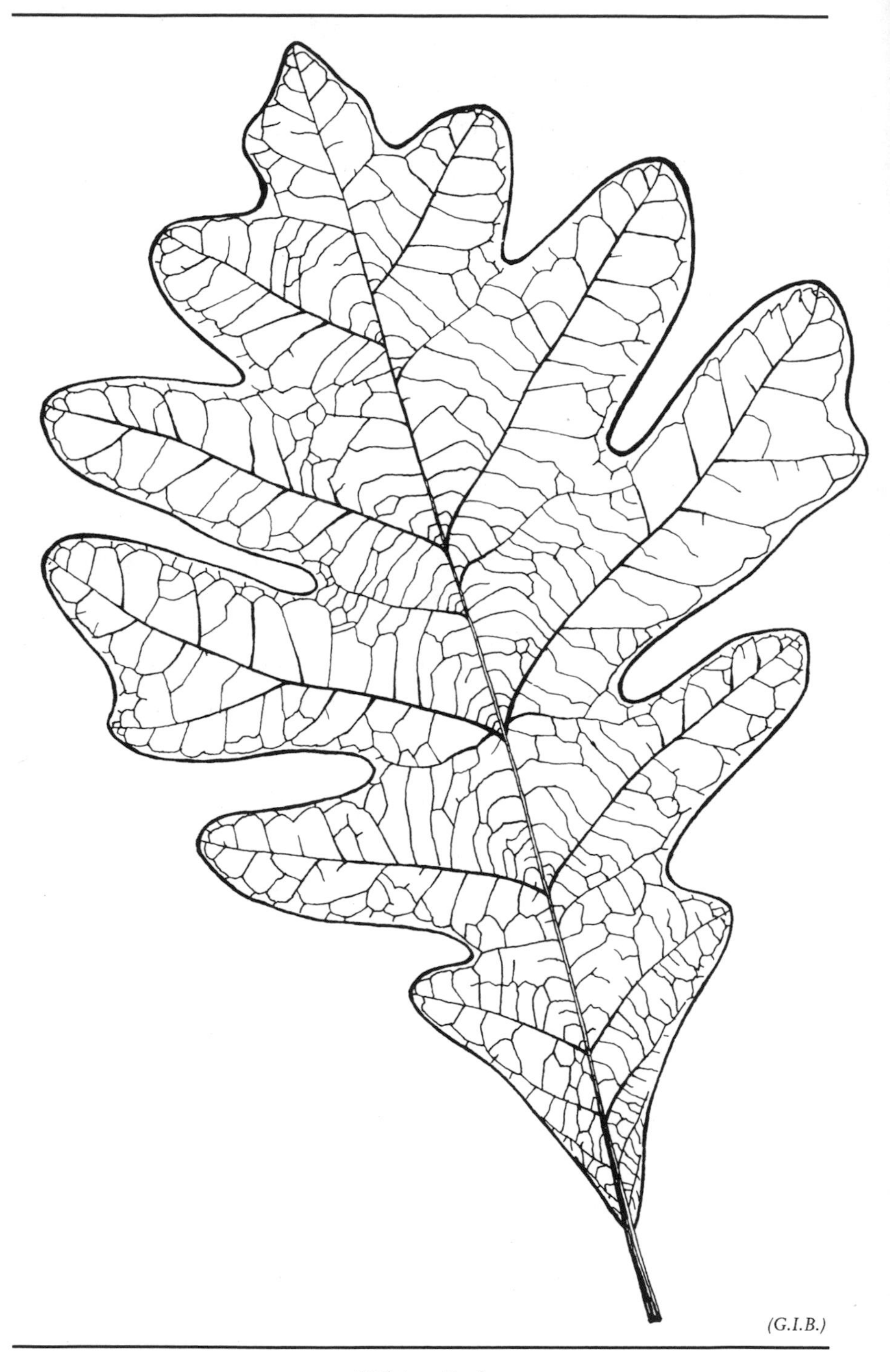

(G.I.B.)

White Oak

White Oak

Quercus alba L.

Range: Eastern U. S. and Southern Canada. In New Hampshire only in southern part.

Habitat: Dry woods, infrequent to common, especially on sandy soil.

Profile: Large tree with spreading crown.

Branching: Alternate.

Leading Shoot: Straight.

Leaves: TIPS OF LOBES SMOOTH, ROUNDED, NO SPINES. Prominent lobes 4–10, *rounded. Smooth underneath,* green lobes variable in depth. Leaves *not broader above* the middle. Leaves reddish to violet in autumn. Brown *dead leaves remain on tree in winter.*

Buds: *Blunt, smooth,* brownish at tips of twigs in clusters.

Twigs: Purplish gray to greenish red, moderately stout, *smooth.*

Bole: Moderate taper, crown almost half of tree length.

Bark: GRAY, thin on young trees. FLAKY, irregularly plated or grooved.

Flowers: Both sexes on the same tree. Male flowers: clusters of hanging beads. Female flowers: few, in angles of new leaves. Greenish, yellowish or reddish.

Fruit: Maturing in *one season.* Acorn cup with wart-like scales. Cup shorter than nut. *Sweet* and edible.

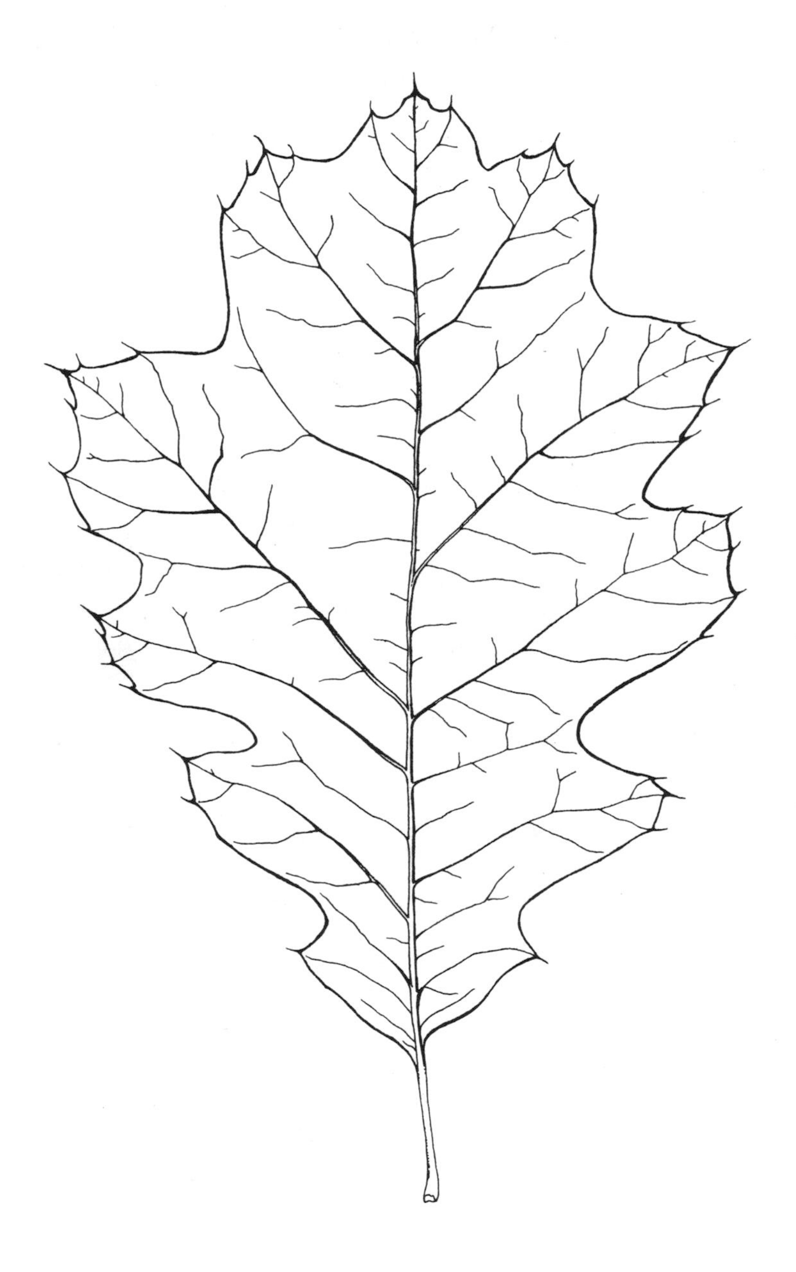

N.T.S. *(G.I.B.)*

Black Oak

Black Oak

(Yellow-Barked Oak)

Quercus Velutina Lam.

Range: Eastern U. S. North to central New Hampshire. Rare in the northern part of its range.

Habitat: Moist, rich, well-drained soils.

Profile: Large tree similar to red oak.

Branching: Alternate.

Leading Shoot: Straight.

Leaves: TIPS OF LOBES, SHARP-POINTED WITH SPIKES ON ENDS OF TEETH. Straight, 7 lobes, *variable,* some deeply cut, others with *straight to incurving sides. Downy or hairy beneath,* rough, yellow-green copper below. Yellow autumn foliage, falling early.

Buds: *Pointed,* 5-sided, large, *densely hairy, gray-brown* to yellow-brown, strongly *angles* in cross section.

Twigs: Stout, reddish brown, *downy or scurfy.*

Bole: Small taper, tall tree, small crown.

Bark: *Dark, rough,* furrowed. INNER BARK ORANGE OR YELLOW.

Flowers: Similar.

Fruit: Maturing in *second autumn* (2 seasons). Cut about 1/2 length of nut. *Upper scales forming a short fringe, bitter* to taste.

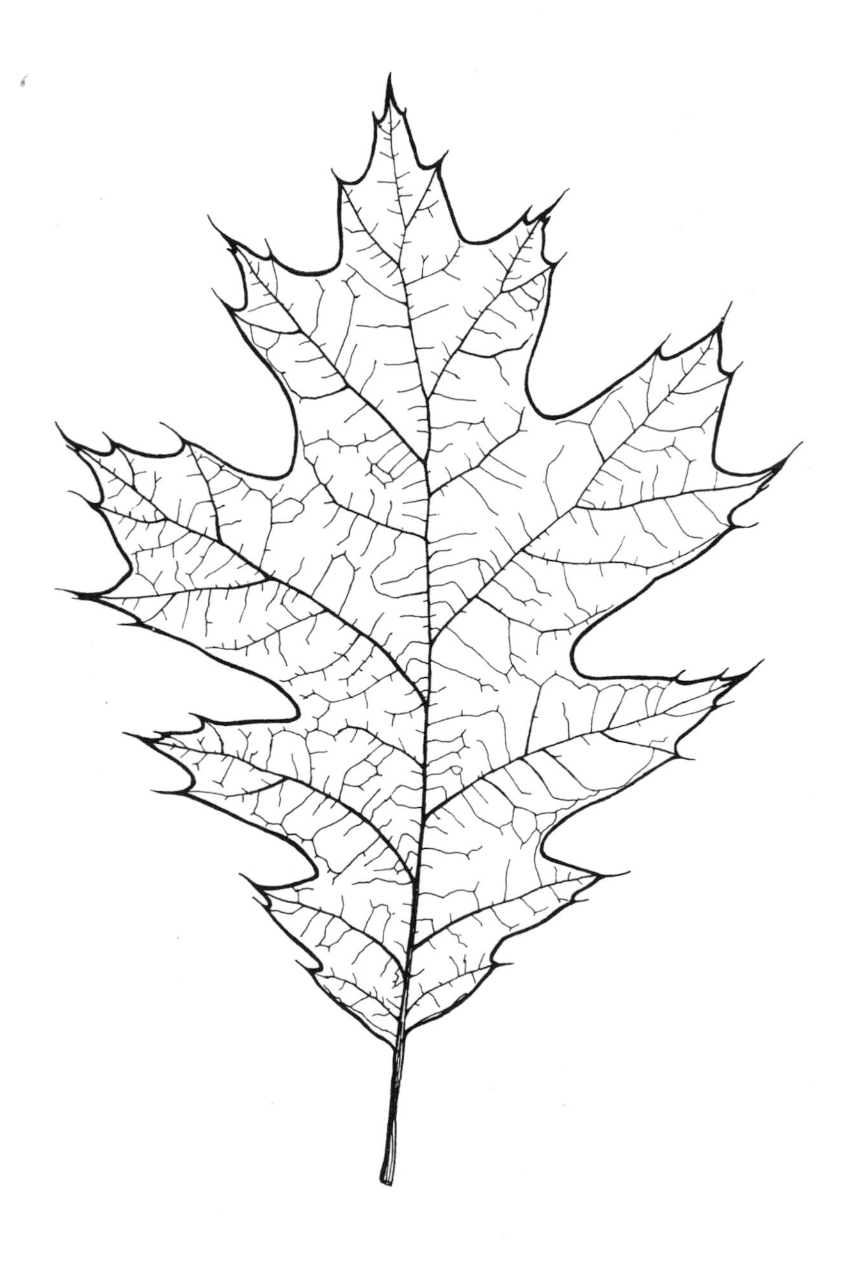

N.T.S. (G.I.B.)

Red Oak

Red Oak

Quercus rubra L.

Range: Eastern U.S. except for the south Atlantic and Gulf Coastal Plains.

Habitat: Sandy loam soils.

Branching: Alternate.

Leading Shoot: Erect.

Leaves: Spikes on tips of lobes. Mostly 3 pairs, lateral lobes, *V-shaped space between* lobes. Lobes 7–11, triangular in outline, *smooth on both sides.*

Buds: Brown, blunt-pointed, smooth, not angled in cross section. *Buds in clusters at ends of twigs.* Larger than other oaks except black. 1/4" (.6 cm) long, rounded on the sides. Terminal buds have slight rusty hairs at the apex.

Twigs: Alternate, strongly ridged or fluted. Reddish brown or black, smooth.

Bole: Largest of the oaks in the Northeast. Straight, clean bole often extends high into crown. Rather round-topped crown.

Bark: Smooth, greenish brown on young stems, becoming broken into ridges, grayish brown, hard with deep furrows. Inner bark reddish, not bitter.

Flowers: Long spreading strings.

Fruit: Acorn maturing in second year. Cup embracing 1/3 to 1/2 the acorn. Ovate to cylindrical. Cup short, shallow. *Large acorns,* bitter, inedible, meat pale yellow.

Scarlet Oak

Scarlet Oak

Quercus coccinea Muenchh.

Range: Eastern U.S. except for the south Atlantic and Gulf Coastal Plains. Rare in southern New England. Southern New Hampshire only.

Habitat: Dry, sandy soils.

Branching: Alternate.

Leading Shoot: Erect.

Leaves: SPIKES on tips of lobes, VERY DEEPLY CUT LEAVES. Sides of lobes incurved. Bright green above, shining, smooth below. Form *rounded spaces between lobes. Red autumn* foliage. Lobes less numerous than red oak.

Buds: Brownish, *somewhat hairy above the middle.* Pointed at tip, gray.

Twigs: Slender, reddish brown, smooth, often orange-red.

Bole: Medium-sized tree, trunk tends to be continuous into the crown.

Bark: Outer bark nearly black and rough, or gray and light brown. Inner bark pale *reddish* or gray.

Flowers: Slender, recurving strings.

Fruit: Acorn maturing in the second year. Ovoid, twice as long as cup. Cup thin, top-shaped, hemispherical. Nut has white meat. Overlapping flat scales do not form a fringe.

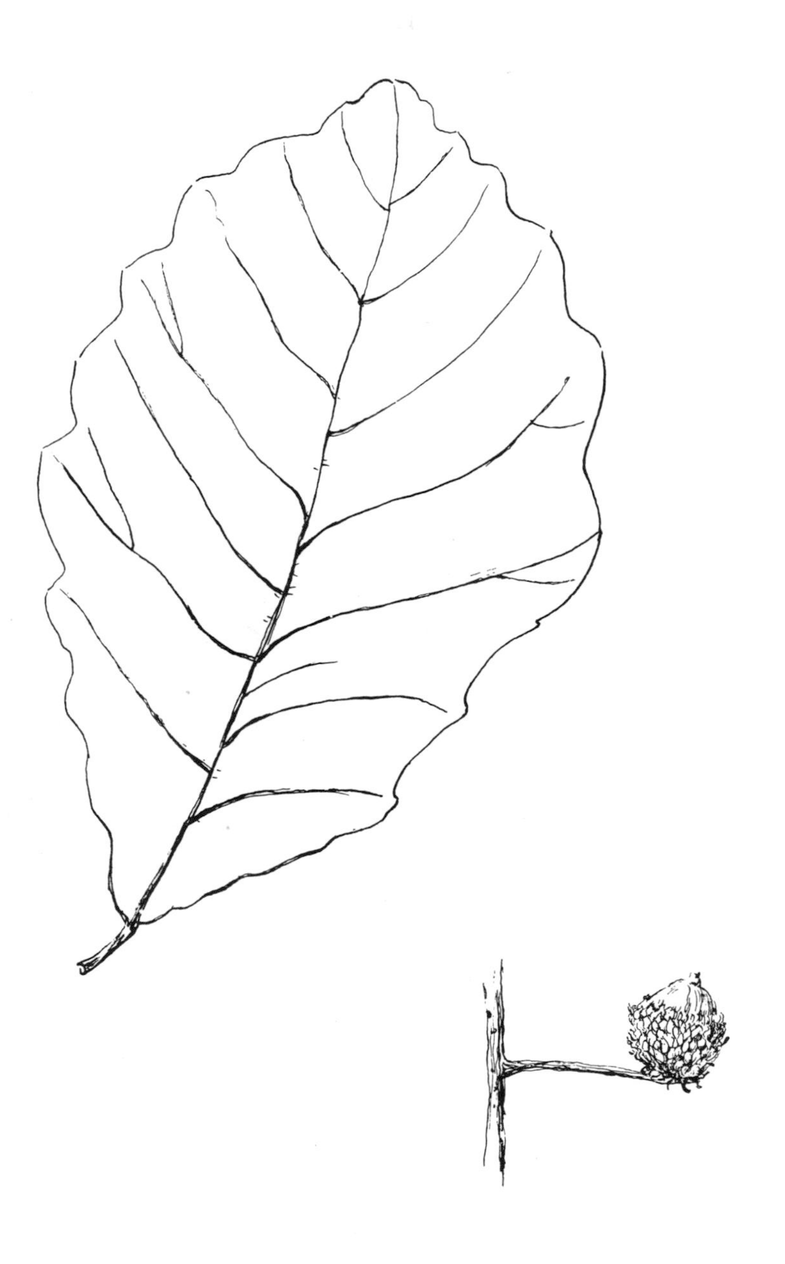

N.T.S. (P.K.)

Swamp White Oak

Swamp White Oak

Quercus bicolor Willd.

Range: South Maine, Quebec, west to Minnesota and New England, south. Infrequent in New England.

Habitat: Stream margins, swamps.

Profile: Irregular crown. Uneven, scraggly tree in the north, becoming large tree in south.

Branching: Alternate.

Leaves: Oval, 5–6" (12–15 cm) long and 2–4" (5–10 cm) broad, *broadest above the middle, strongly tapered* toward the *wedge-shaped* base. Margin shallowly lobed. Downy or white beneath.

Buds: Spherical, terminal buds orange-brown, *blunt-pointed,* smooth, lateral buds crowded, small.

Twigs: Slender to stout, straw-brown to yellow-green. *Bark peels and scales from twig* (only oak where this occurs).

Bark: *Flaky, peeling in ragged papery scales* (resembles the manner in which sycamore bark peels). Furrowed at base of tree in long, flat scaly ridges.

Flowers: Male and female borne separately on the same tree, the male in catkins; female in short spikes.

Fruit: Acorn cup *1/3 to 1/2 as long as* acorn. Cup does not half cover the nut. Upper scales of cup pointed, often forming a fringe.

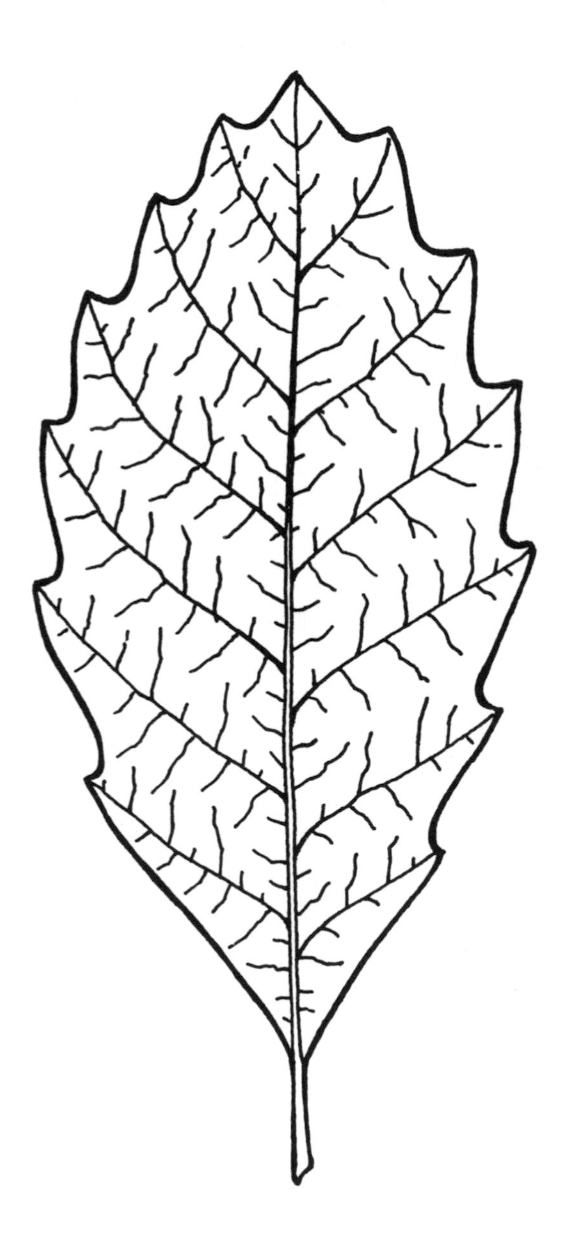

(G.I.B.)

Scrub or Dwarf Chestnut Oak

Scrub or Dwarf Chestnut Oak

Quercus prinoides Willd.

Range: Southwest Maine, west to Minnesota, south to Alabama.

Habitat: Dry, sandy, rocky ground.

Profile: Slender shrub, 3–11' (1–3 m), rarely treelike.

Branching: Alternate.

Leaves: Oblong or inversely egg-shaped, *not deeply lobed.* 3–5" (7–12 cm) long. Margins wavy with 3 to 7 blunt teeth on each side. Bright green above, whitish beneath with dense white hairs.

Buds: Oval, small, *very blunt,* clustered at end of twig. Terminal bud woolly.

Twigs: Slender, smooth, light orange brown, brittle.

Bark: Thin, light brown to ashy, *scaly.* Begins to *scale off when* stem is 1 1/2" in diameter.

Flowers: Male and female borne separately on the same tree, male in catkins, female in short spikes.

Fruit: Short stalked; acorn cup 1/3 to 1/2 as long as nut. Cup has wart-like scales.

(P.K.)

Black Scrub Oak

Black Scrub Oak

Quercus ilicifolia Wang.

Range: New England to South Carolina.

Habitat: Sandy barrens.

Profile: Shrub or low tree to 20' (6 m) with stiff, twisted branches.

Branching: Alternate.

Leaves: *Deeply lobed,* 2–4" (5–10 cm) long with 3 to 7, usually 5 bristle-tipped lobes. The curves of the lobes are broad and shallow. Dark green and shiny above, the lower side whitened by a pale down.

Buds: Short, *sharp-pointed,* chestnut brown, *shiny.*

Twigs: Slender, gray brown, or olive, *slightly downy, the first* season, becoming dark and smooth.

Bark: Dark, greenish to grayish brown, smooth, *never breaking into scales.*

Flowers: As in other oaks.

Fruit: Short stalked. Acorns 1/2" (1 cm) high, cups saucer-shaped downy on the inside.

(G.I.B.)

American Elm

American Elm

(White Elm, Water Elm)

Ulmus americana L.

Range: South Canada to Florida and west to Texas.

Habitat: Common inhabitant of wet flats where standing water may accumulate in the spring and fall.

Profile: Spreading, vase-shaped, large tree.

Branching: Alternate.

Leaves: *Lopsided at base* with prominent parallel veins, slightly toothed *rough on upper surface, smooth* to hairy below. Large 4–6" (10–15 cm), light reddish brown, shiny.

Buds: *Smooth* or somewhat pale, small, red-brown, a little to the side of leaf scar, terminal bud absent.

Twigs: SMOOTH, slender, brown somewhat zigzag, lenticels pale, scattered, inconspicuous .

Bark: Grayish brown, rough, furrowed. Fibrous layers separated by *light-colored corky patches* when cut.

Bole: Wood hard to split, branch early.

Branches: Drooping.

Flowers: Reddish green, drooping clusters appearing before the leaves. *Stalked.*

Fruit: Hairy on the margins, winged seeds *small, with hairs, long-stalked* clusters. Ripening in May. Deeply notched at the apex.

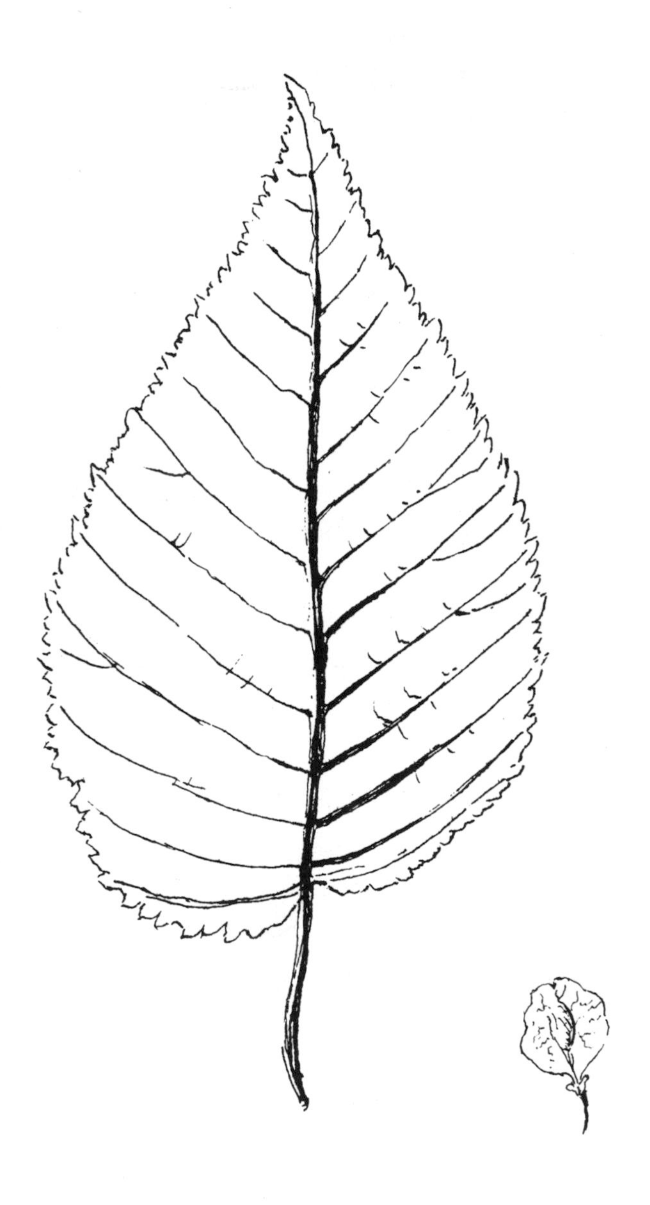

(P.K.)

Slippery Elm

Slippery Elm

(Red Elm, Moose Elm)

Ulmus rubra Muhl. (U. *fulva* Michx.)

Range: South Canada south to Florida and Texas, west to North Dakota, rare in New England.

Habitat: Prefers limestone areas.

Profile: Small to medium-sized tree, less spreading .

Branching: Alternate.

Leaves: *Lopsided at base,* toothed, *very rough* on top, *soft-downy underneath,* large 2–6" (5–15 cm). *Fragrant when dry.*

Buds: Dark brown, *large,* nearly black. Winter buds *downy with rusty hairs.* Terminal bud absent.

Twigs: HAIRY or WOOLLY, roughened by numerous raised bumps. *Strongly mucilaginous* when chewed.

Bark: Grayish brown, internally reddish brown. Inner bark next the wood whitish, *strongly mucilaginous. No white layers in outer bark.*

Bole: Wood easy to split. Long clear length.

Branches: Tend to point upward.

Flowers: Flowers *without stalks,* small yellow appearing in April.

Fruit: Wide-winged, very large, round, *without hairs,* borne in *short*-stalked dense clusters. Ripening in May; smooth, thin, papery.

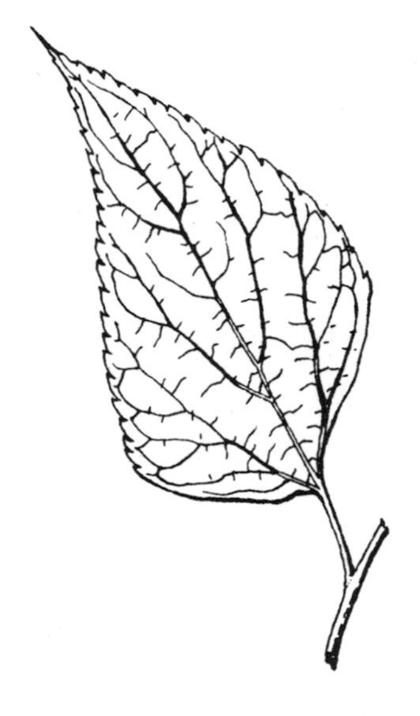

(G.I.B.)

Hackberry

Hackberry

Celtis occidentalis L.

Range: Southern New Hampshire, Vermont, south to Florida.

Habitat: River banks and rich woods. Often in limestone areas.

Profile: Large tree or coarse shrub depending on site. A short bole much divided to form an oblong, rounded crown.

Branching: Alternate.

Leaves: Narrow, oval, 2–5" (5–12 cm) *lopsided at base,* toothed on at least the broader side (10–40 teeth) with a *long curved tip. Unequally heart-shapted at base.* Rough on upper surface.

Buds: Terminal bud lacking; lateral buds *close-pressed* to stem.

Twigs: Slender, *zigzag,* reddish brown.

Flowers: Greenish, small, on drooping stems, opening with the leaves.

Fruit: Thin fleshy, sweet berries on long *slender stems like cherries.* Purple-black to brown or buff, with a short thick beak. Surface of seed *(pit) with a net-like pattern.*

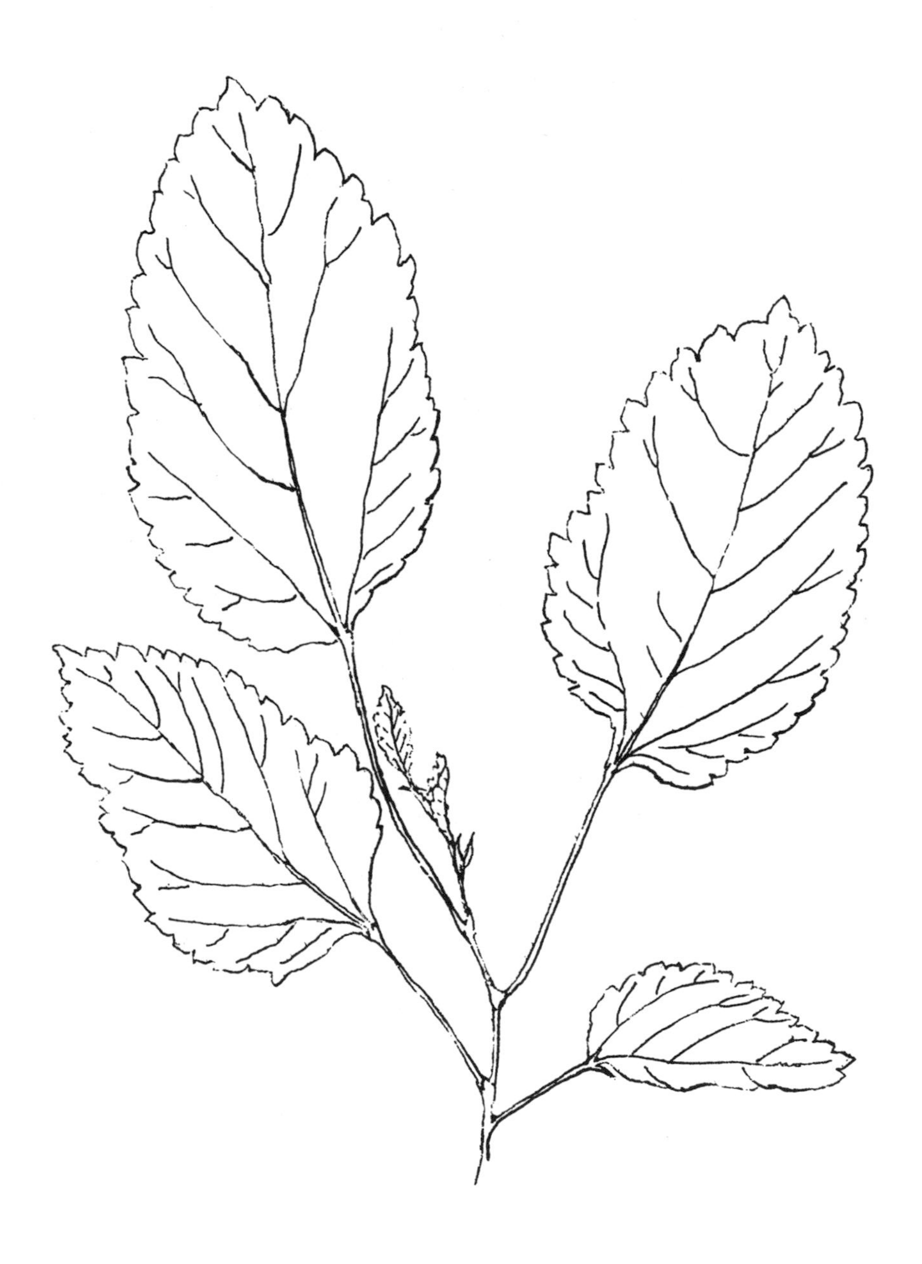

(P.K.)

Red Mulberry

Red Mulberry

Morus rubra L.

Range: Southwest Vermont to Florida, west to Lake states.

Habitat: Bottomlands, rich woods.

Profile: Small or medium sized tree, with a broad rounded top crown.

Branching: Alternate.

Leaves: *Leaf base on a slant.* Coarse 2–4" long (5–10 cm), toothed, pointed on *young shoots, often lobed with milky juice,* rough above, downy beneath.

Buds: Terminal bud lacking. Lateral buds with a *green tinge.*

Twigs: Slender, brown.

Flowers: Sometimes male and female flowers are on separate trees. Male flowers in catkins, female in hanging spikes.

Fruit: Dark purple or red berry resembling a raspberry, edible, ripens in late June or July. Eaten by birds.

(P.K.)

Common Barberry

Common Barberry

Berberis vulgaris L.

Range: Introduced from Europe, but naturalized and common in eastern New England.

Habitat: Sandy loam soils.

Profile: Small, upright shrub 3–9' (1–3 m).

Branching: Alternate.

Leaves: Small, about 1" (2.5 cm) long, in clusters, finely toothed *with bristle at end.* Veins prominent beneath arranged in whorled lateral clusters.

Buds: Small, brownish, 5–6 exposed scales.

Twigs: Branches of second year gray, *with scattered sharp spines,* often in threes.

Bark: *Inner bark yellow; wood yellow.*

Flowers: Yellow, in drooping clusters, 6 petals with smooth edges, 10–20 flowers on stalks.

Fruit: Narrow, red berry less than 1/2" long, oval shaped in a long cluster.

N.T.S. (G.I.B.)

White Sassafras

White Sassafras

Sassafras albidum (Nutt.) Nees

Range: Southern Maine, New Hampshire, Vermont; south to Virginia, west to Michigan.

Habitat: Low ground.

Profile: Small to medium-sized tree.

Branching: Alternate.

Leaves: Variable, some ovate, some 3-lobed and *some mitten-shaped.* Smooth, no teeth, somewhat gummy.

Buds: Green, tinged with red, with *strong odor and taste.* Terminal bud large, with 4 scales.

Twigs: Pale green, smooth when young, downy or gummy, rapidly growing, surpassing the main stem in length during the first season.

Bark: Old bark *spicy-aromatic* deeply furrowed and with narrow horizontal cracks. *Green when young.*

Bole: Branchy.

Flowers: Male and female flowers on separate trees. Inconspicuous yellowish green in clusters, appearing with the leaves in April.

Fruit: Clusters of dark blue, round berries on fleshy red stalks. The scaly capsules ripen in September, each with a single seed.

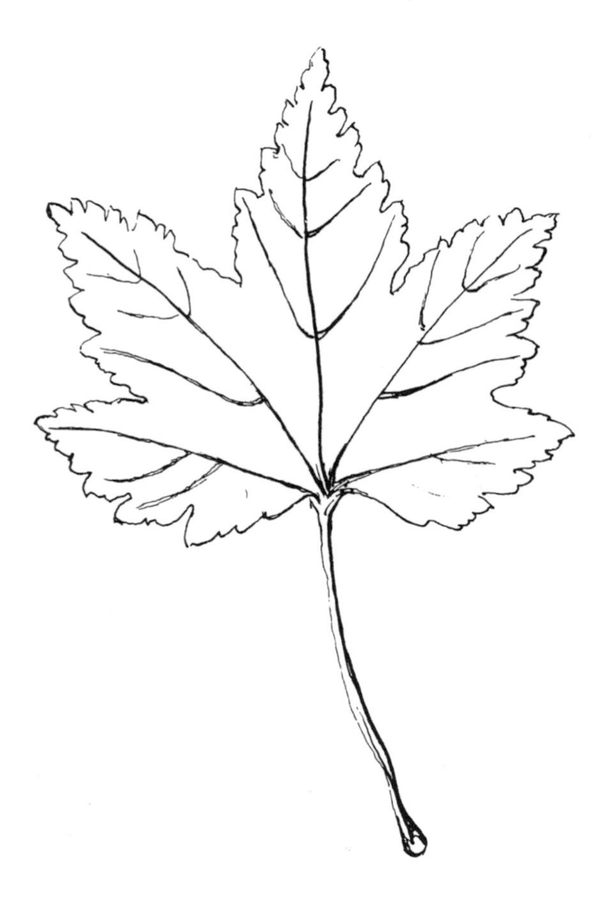

(P.K.)

Skunk Currant

Skunk Currant

Ribes glandulosum Grauer

Range: Frequent in northern New England.

Habitat: Often in damp woodlands and alpine areas.

Profile: Trailing or erect shrub to 2' (5 m) high.

Branching: Alternate.

Leaves: Rounded, ovate 3–7 lobed and toothed. *Odor of skunk when crushed*, one of the *first green leaves* to appear in spring.

Buds: Greenish to *purple-red.*

Twigs: *Smooth* with no bristles or prickles, slender, slightly zigzag, pale olive brown. *Skunk odor when crushed.*

Bark: Smooth, almost satiny, dark brown to black.

Flowers: White to pink, early May.

Fruit: *Red berry 1/4" (0.6 cm) long*, covered with small hairs. Edible.

(G.I.B.)

Prickly Gooseberry

Prickly Gooseberry

Ribes cynosbati L.

Range: Scattered in northern New England.

Habitat: Dry, rocky woodland.

Profile: Small shrub up to 4' (1.5 m).

Branching: Alternate.

Leaves: 1–2" (2–5 cm) long, lobed and toothed, slightly hairy beneath on long thin stalks, *no odor.*

Buds: Pointed, *pale brown and hairy.*

Twigs: Covered with scattered THIN SPINES ABOUT 1/4" long, dull grayish brown, *no odor.*

Flowers: Small, green in May.

Fruit: Berry, 1/2" (1 cm) in diameter, ripening in July with LONG STIFF PRICKLES, *wine-colored.* Edible.

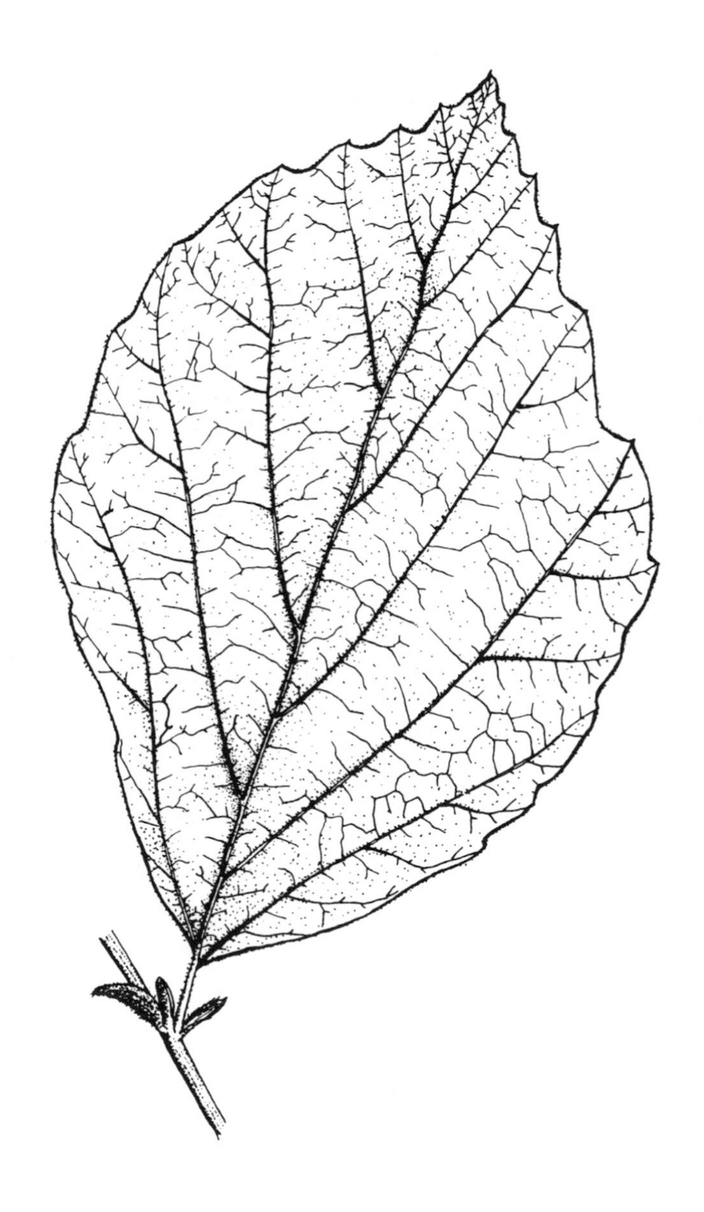

(G.I.B.)

Witch Hazel

Witch Hazel

Hamamelis virginiana L.

Range: Southern Canada, south to Georgia, Missouri.

Habitat: Dry to moist woods, *deep shade*. Does NOT sprout.

Profile: Large shrub, 8–10' (3 m).

Branching: Alternate.

Leading Shoot: Curving, zigzag, short shoots.

Leaves: *Variable, obliquely rounded, wavy, lopsided,* oval, toothed in shallow rounded lobes. Main veins prominent, 5–7 nerves. Green both sides, one-sided at base.

Buds: Flattened, with rusty brown hairs. Terminal bud stalked, curved or sickle-shaped.

Twigs: Zigzag, smooth to slightly hairy. Short shoots light orange brown. Some brown warts, but lenticels indistinct.

Bark: Mottled, brown, smooth to scaly.

Flowers: Bright yellow, opening *in October–November while fruit of year still present.* Crinkle, curving *petals, persisting to winter,* remaining attached to twig.

Fruit: A woody, hairy 2-seeded pod, splitting open and *ejecting black seeds forcibly, ripening a year after flowers.* Open pods remaining on the stem.

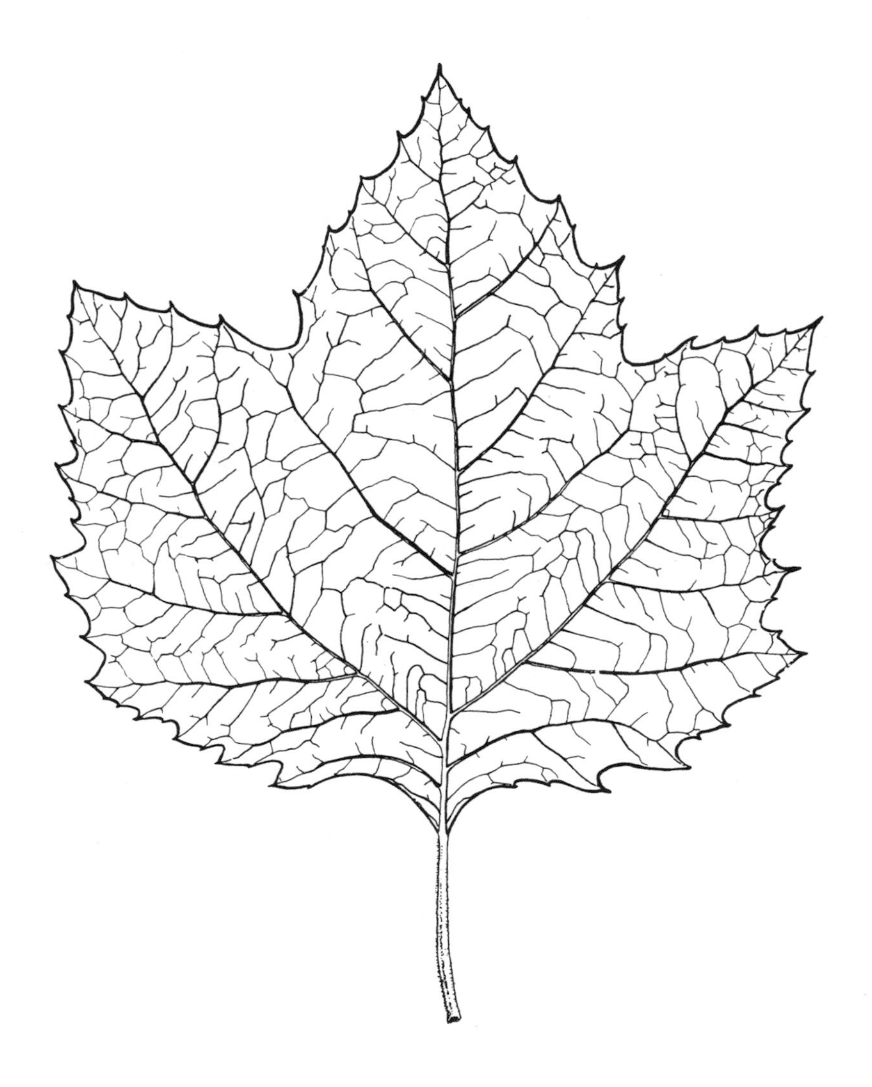

N.T.S. (G.I.B.)

Sycamore

Sycamore

(Plane Tree, Buttonwood)

Platanus occidentalis L.

Range: Southern New England along Connecticut River, southward and west to Kansas and Texas.

Habitat: Stream banks and bottomland.

Profile: Large tree, perhaps largest tree in the east.

Branching: Alternate.

Leading Shoot: Somewhat crooked.

Leaves: Large, broad, oval with 3–5 lobes, maple-like, hairy beneath, with long pointed teeth.

Buds: Long, conical, resinous, shiny, dark, red-brown, covered by a single cap-like scale. No terminal bud.

Twigs: Slender, smooth, yellow-brown, zigzag, swollen at joints. Pith thick.

Branches: Ascending or horizontal spreading.

Bole: Gradual taper.

Bark: *Creamy white on young branches,* turning red-brown-gray, *then breaking in large thin plates exposing* WHITISH INNER BARK, giving a white-washed appearance.

Flowers: Small, inconspicuous.

Fruit: Large round balls 1" (2.5 cm) in diameter, with bristly hairs at base, containing many seeds. Borne SINGLY ON LONG STALKS. Remains on tree until spring.

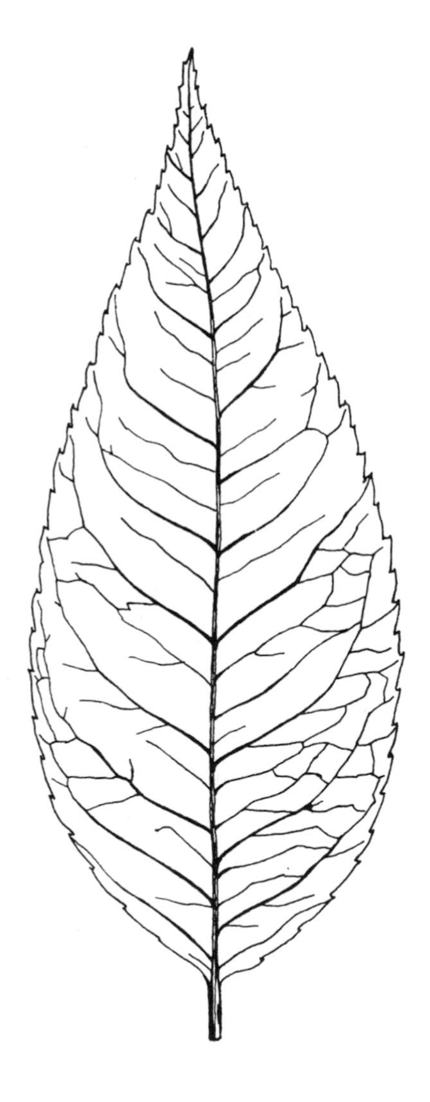

(G.I.B.)

Black Cherry

Black Cherry

(Rum Cherry)

Prunus serotina Ehrh.

Range: Eastern U.S. and Canada.

Habitat: Rich woods.

Profile: Large tree with spreading crown.

Branching: Alternate.

Leading Shoot: Long, straight, wand-like.

Leaves: 4" (10 cm) long, 1" (2.5 cm) wide. Finely toothed. Smooth above, shiny, teeth turning inward. BROWN WOOL ON BASAL 1/3 of MIDRIB ON UNDERSIDE. Yellow to brown in autumn.

Buds: Bright, light reddish brown, 4 scales appear as none. Terminal bud *larger* than laterals.

Twigs: *Very rancid tasting and evil smelling.* Often has black knot swellings on branches, elongated lines or lenticels.

Bole: Straight in forest trees, crooked in open-grown trees.

Branches: Horizontal.

Bark: Dark red brown, smooth on young trees, becoming roughened with dark scales on older trees. Inner bark rancid smelling and tasting. Narrow, horizontal lenticels. Appearance resembles black birch.

Flowers: White, elongated clusters, bead like, 4" (10 cm) long, when leaves are 1/2 grown.

Fruit: 1/3 to 1" (.5–2 cm) diameter, dark red, becoming black with small stone. June–October. Flesh dark purple. Edible.

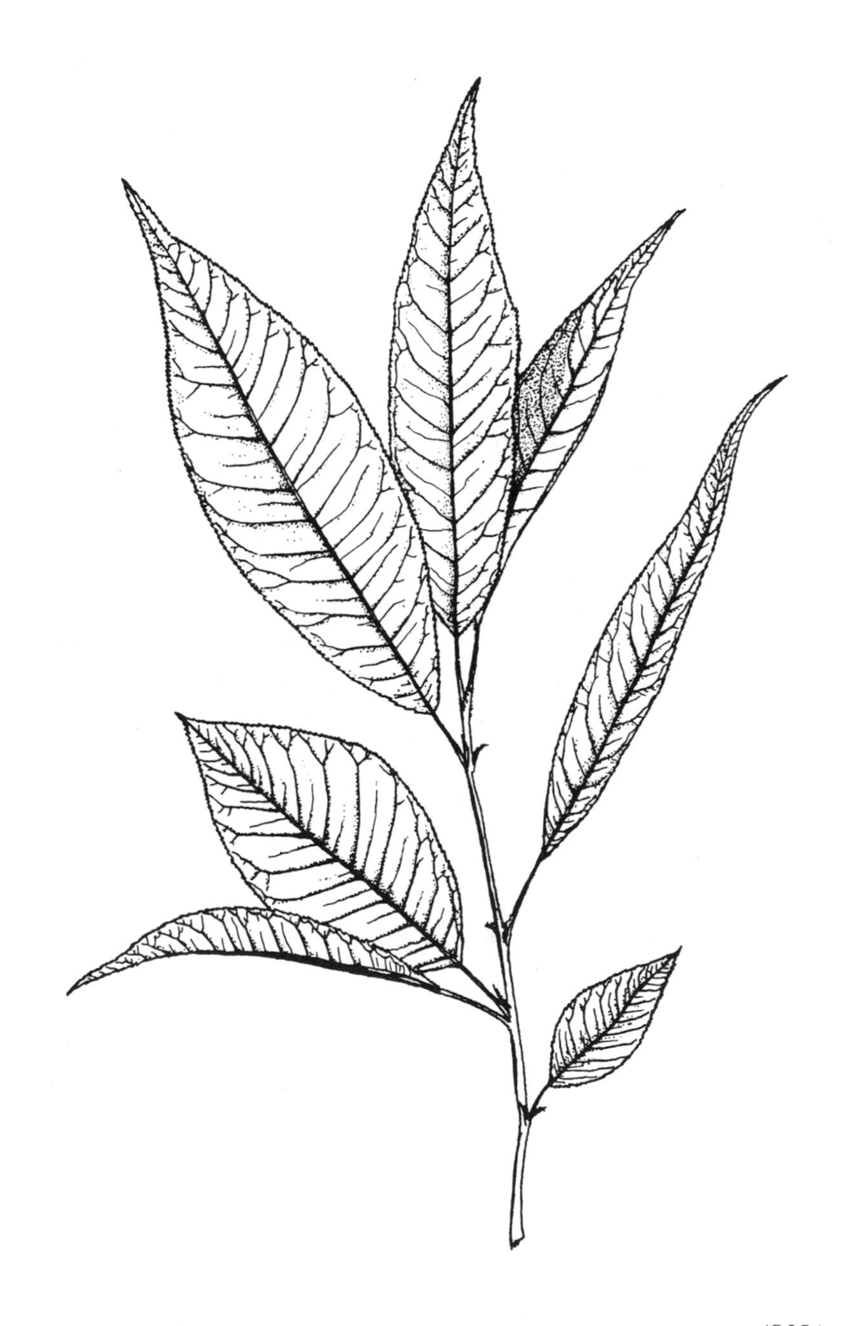

(G.I.B.)

Pin Cherry

Pin Cherry

(Fire Cherry, Bird Cherry)

Prunus pensylvanica L.f.

Range: Northern U.S. and Canada.

Habitat: Especially after fires and cutting.

Profile: Small slender tree.

Branching: Alternate.

Leading Shoot: Slender, straight.

Leaves: VERY NARROW, LONG, finely toothed. *Shining smooth on both sides.*

Buds: Terminal bud *smaller* than associates. Minute, blunt pointed, often covered by a grayish skin. Clustered at the ends of fruiting spurs.

Twigs: Only *slightly rancid-tasting* or evil smelling. Thin, reddish.

Bole: Slender, straight.

Branches: Ascending.

Bark: *Red, thin, somewhat papery,* bitter tasting.

Flowers: White clusters, expanding with the leaves. Petals round, 1/2" (1 cm) broad, all arising from same point on the stem.

Fruit: *Red, sour,* with large *stone,* thin acid flesh.

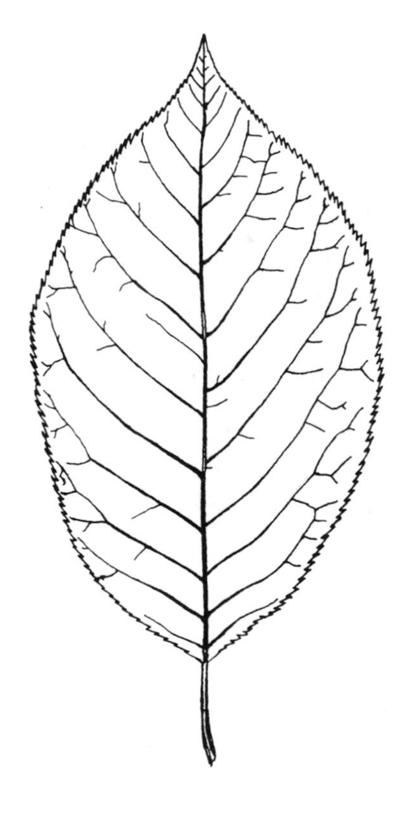

(G.I.B.)

Choke Cherry

Choke Cherry

Prunus virginiana L.

Range: Eastern U.S. and Canada

Habitat: With aspens in burned areas; dry soils.

Profile: Shrub or sometimes small tree with many stems.

Branching: Alternate.

Leading Shoot: Straight.

Leaves: Oval with sharp teeth. BROADEST ABOVE MIDDLE. Broader and more oval than black cherry. No down beneath.

Buds: Small, thin, tapering to less than 1/2" (1 cm). Bud scales roughened.

Twigs: VERY RANCID TASTE AND SMELL. Stout, coarser than black cherry, reddish to grayish brown, smooth. *No grayish skin that can be rubbed off.*

Bole: Many-stemmed, curving.

Bark: Smooth, gray brown on young trees; easily peeled off in dark papery layers. Does NOT become rough or scaly with age.

Flowers: White, small 1/8" (4 mm) long, clusters terminating the seasons growth.

Fruit: Red berry, becoming purple; acid, astringent, puckery taste.

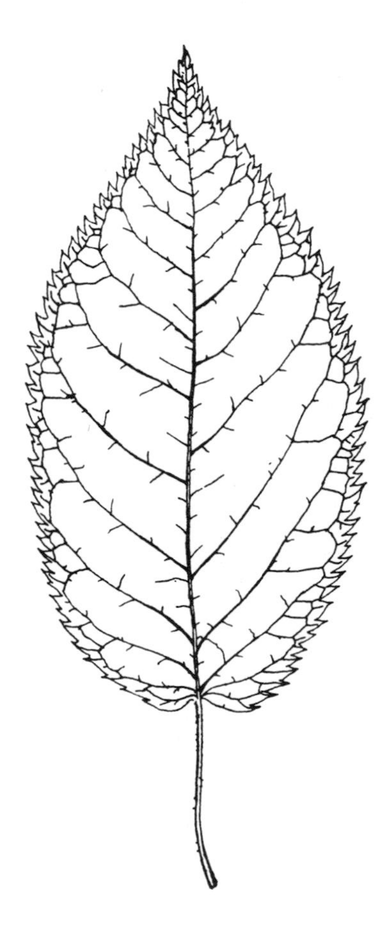

(G.I.B.)

Shadbush

Shadbush

(Service Berry, June Berry)

Amelanchier canadensis (L.) Medic.

Range: Eastern U.S. and Canada.

Habitat: Sandy, moist, well-drained soils.

Profile: Small tree or shrub, few stems. Vase-like clumps of stems.

Branching: Alternate.

Leading Shoot: Straight.

Leaves: Oval, finely toothed, broadest above the middle. Veins curved. When half grown with *white felt beneath or woolly hairs.*

Buds: Terminal buds long, narrow; greenish yellow or purple.

Twigs: NO RANCID TASTE OR SMELL (slight taste of bitter almonds). *Slender, grayish, often covered by a gray skin, smooth.* Lenticels are scattered minute dots.

Bole: Often straight tree 40' (12 m).

Bark: GRAY, smooth, older trees with vertical fissures, separating smooth flat areas.

Flowers: White cluster appearing when leaves are $1/2$ grown.

Fruit: *Red becoming purple black. Highly edible.*

(G.I.B.)

Meadowsweet

Meadowsweet

Spiraea latifolia (Ait.) Borkh.

Range: Northeastern Canada, west to Michigan.

Habitat: Pastures and low ground.

Profile: Low shrub.

Branching: Alternate.

Leaves: Thin, narrow, coarsely toothed above the middle 2–3" (5–8 cm) long, blunt, smooth, short stalked.

Buds: Small, numerous, smooth.

Twigs: *Smooth,* slender, light reddish brown to purple; more or less angled.

Bark: Thin, peeling in filmy layers.

Flowers: Numerous, very small, *white or pale,* pink, smooth. In conical clusters. June.

Fruit: Small, hard, *smooth,* brown pods in groups of 5 on each stalk, forming pyramidal clusters.

Steeplebush, Hardhack

Steeplebush, Hardhack

Spiraea tomentosa L.

Range: Northeastern Canada, south to North Carolina.

Habitat: Pastures and sterile ground,

Profile: Low shrub.

Branching: Alternate.

Leaves: Pointed at base, 1–2" (2–5 cm) upper surface bright green. LOWER SURFACE VERY WOOLLY, *with brownish, rusty or tawny hairs.*

Buds: *Buds covered with wool.* No terminal bud.

Twigs: VERY WOOLLY, slender, purple-brown.

Bark: Thin, hairy, peeling.

Flowers: *Deep pink or rose*-colored, very crowded in dense clusters. July–September.

Fruit: Small, *downy,* pale brown.

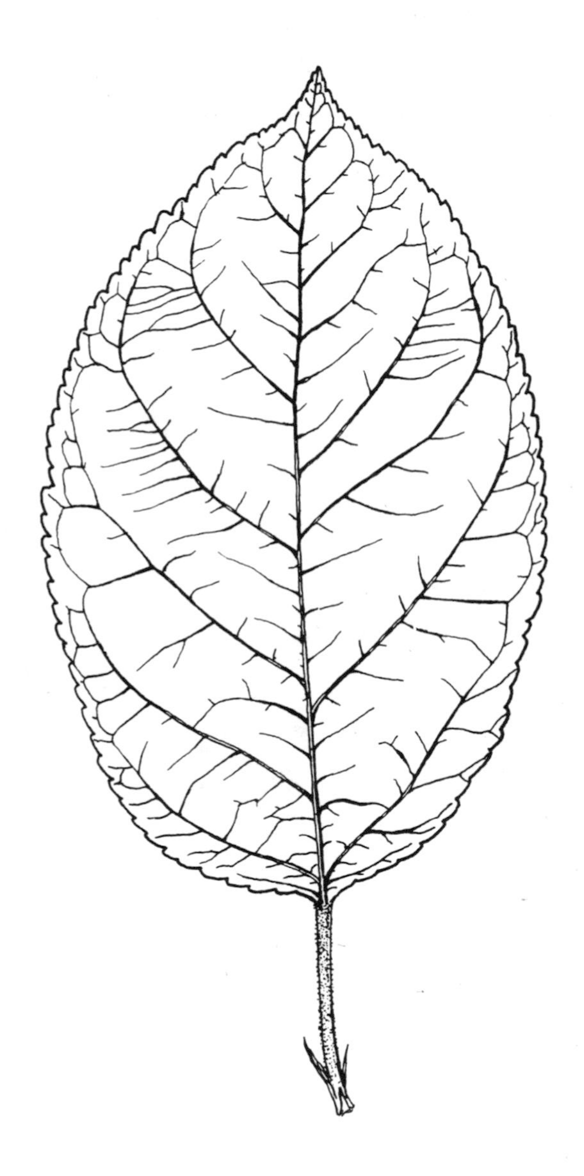

(G.I.B.)

Apple

Apple

Pyrus malus L.

Range:	Eastern U.S. and Canada. Introduced from Eurasia.
Habitat:	Common on overgrown pastures and young forest. Moist, sandy loam, well-drained soil.
Profile:	Spreading low tree up to 30' (10 m) in the forest.
Branching:	Alternate.
Leading Shoot:	Short, straight.
Leaves:	Oval, finely toothed, hairy beneath.
Buds:	Grayish, hairy at tip, blunt terminal bud larger than laterals.
Twigs:	Branches with *short, hard stems but no true thorns.*
Branches:	Curling, twisting.
Bole:	Straight but branching near the ground.
Bark:	Gray and flaking.
Flowers:	White to pink, numerous.
Fruit:	Fleshy, round, containing 5 or more dark, hard seeds. Edible.

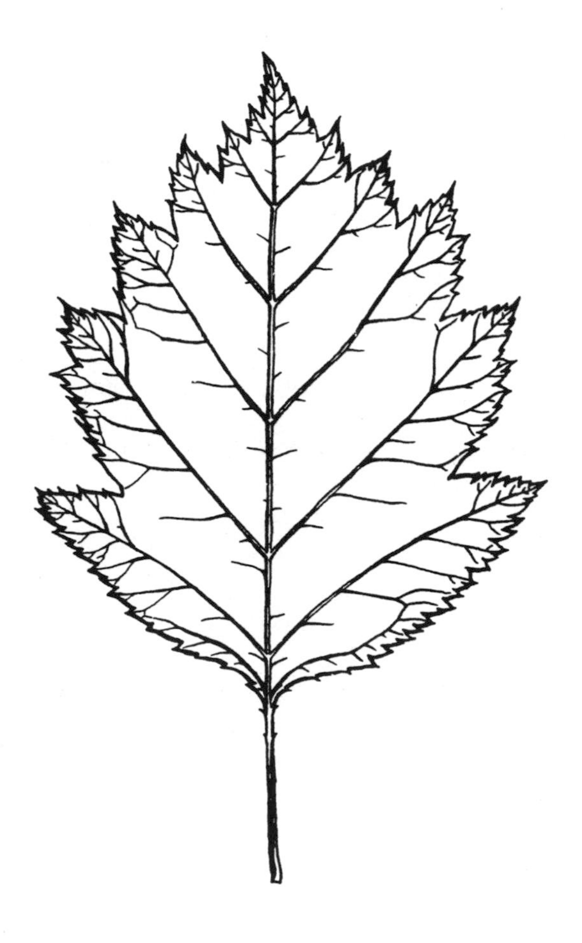

(G.I.B.)

Hawthorn

Hawthorn

(Thorn Apple)

Crataegus Sp.

Range: Central and eastern U.S. and Canada.

Habitat: Open pastures; sandy loam, well-drained soils.

Profile: Shrub, round-crowned, sometimes small trees.

Branching: Alternate.

Leading Shoot: Crooked

Leaves: *Toothed and lobed,* of VARIOUS SHAPES. Leaves at ends of *shoots more deeply cut.*

Buds: Small, rounded.

Twigs: Thorns in the axils (notches) behind the leaves. *Often very thorny* with LONG, VERY STRAIGHT AND VERY HARD SPINES.

Branches: Dense, curving, intertwined.

Bole: Many-stemmed or branching at ground.

Bark: Light gray, smooth.

Flowers: White, occasionally pinkish.

Fruit: Small, apple-like fruits. Inedible.

Remarks: Many species and hybrids.

(P.K.)

Black Chokeberry

Black Chokeberry

Pyrus melanocarpa (Michx.) Willd.

Range: Northern New England.

Habitat: Usually moist or swampy locations. Small shrub, 3–6' (1–2 m).

Branching: Alternate.

Leading Shoot: Rather straight.

Leaves: Ovate, smooth, 1–3" (2–8 cm) upper surface with *small blackish glands along the pale midrib.* Pointed to blunt tip, finely toothed margins. Lower surface paler, often woolly; leaf stems smooth; autumn foliage bright red.

Buds: Oblong, pointed, purplish red with 3–5 exposed scales.

Twigs: Slender, straight, smooth, yellowish brown.

Flowers: White in clusters with 5 petals, 3/8" (1 cm) across. June.

Fruit: Stalks *without hairs*. Round purple to black, 1/4" (.6 cm), September–October. Often persistent into winter. Open clusters.

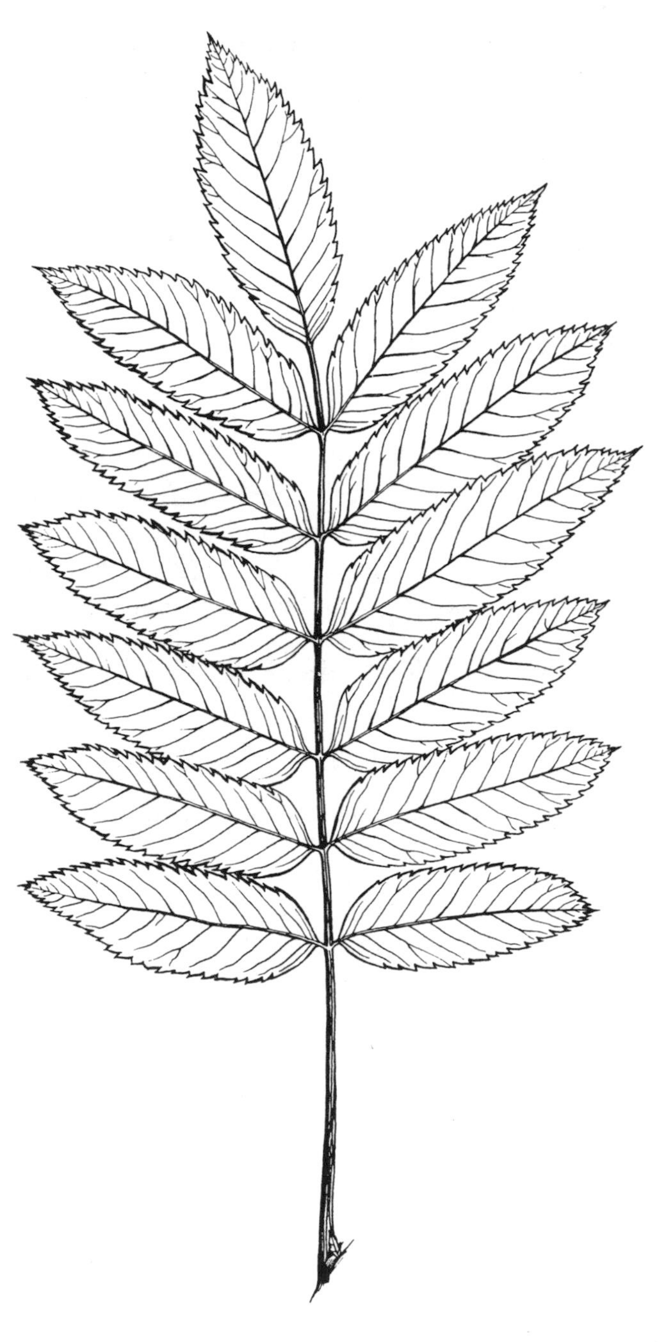

N.T.S. (G.I.B.)

American Mountain Ash

American Mountain Ash

(Round Wood, Rowan Tree, Dogberry)

Pyrus americana

Range: Newfoundland west to Minnesota, south to Georgia, east to Tennessee.

Habitat: Upland forests.

Profile: Small tree or shrub.

Branching: Alternate.

Leading Shoot: Straight.

Leaves: Compound, 11–17 leaflets 6–12" long (4–10 cm), narrow, *3 1/2–5 times as long as broad.* Leaflets *sharp-toothed,* light green above, pale, woolly below when young. Leaves about 10" (25 cm) long.

Buds: Shiny, pointed 1/2" (1 cm) long, *covered by gum. Terminal buds large,* dark purplish red.

Bole: Straight, much branched.

Bark: Thin, smooth with small plate-like scales.

Flowers: 1/4" (5–6 mm), broad. Small, white in large flat clusters, *very numerous.*

Fruit: Bright red berry 1/4" (.5 cm) in diameter.

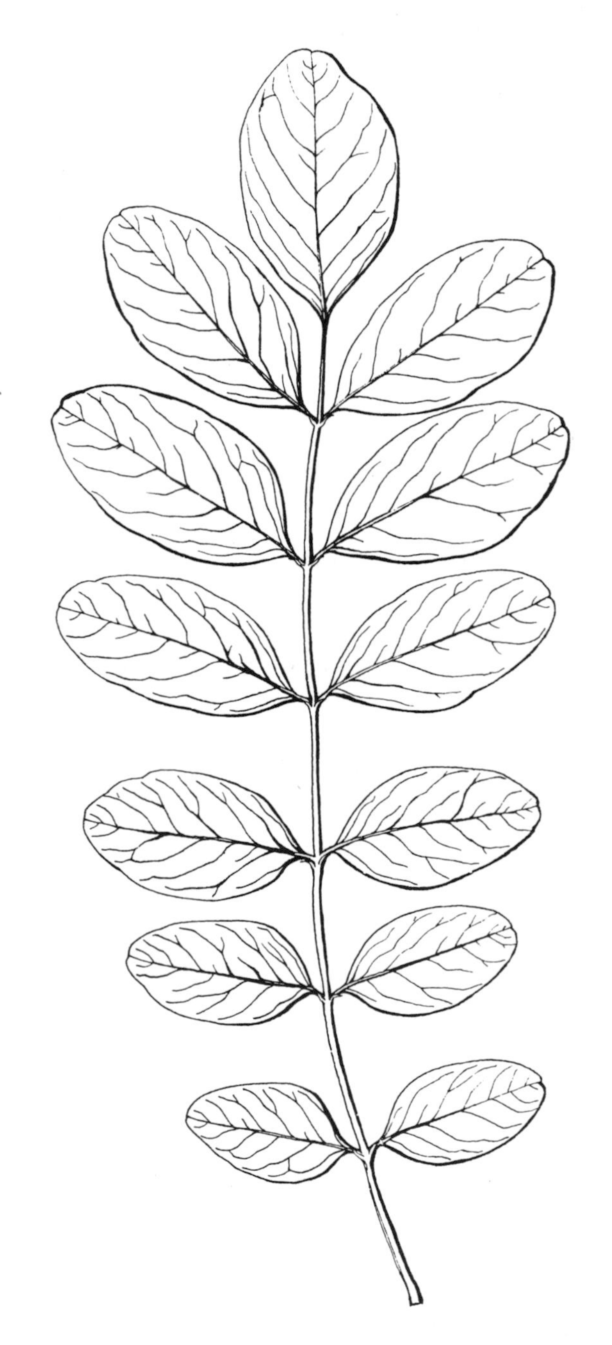

N.T.S. (G.I.B.)

Black Locust

Black Locust

(False Acacia)

Robinia pseudoacacia L.

Range: Central and southern U.S. Escaped from cultivation and naturalized in New England.

Habitat: Rich woods.

Profile: Large tree

Branching: Alternate.

Leading Shoot: Straight to crooked.

Leaves: Compound 7–19 leaflets. Leaflets *rounded, notched at apex*. 1½–2" (4–5 cm) long *without teeth*. Leaflets with whisker-like tip. Leaf about 10" (25 cm) long.

Buds: *Terminal bud lacking*. Side buds small, sharp-pointed, in the *angle by a thorn,* rusty-downy, crowded.

Twigs: Slender, brittle, often *zigzag,* dull brown with many small dots or warts. SHARP SPINES ½–1" (1–2 cm). Pith light brown.

Branches: *Thorny,* dark brown.

Bole: Usually with wavy sweep. Lower branches die early.

Bark: Deeply furrowed brown with *interlacing fibrous ridges. Inner bark green.* HEARTWOOD green.

Flowers: White, *fragrant,* in large clusters 1" (2.5 cm) long.

Fruit: Flat pod, 3" (8 cm) long containing hard, dark brown seeds; bony.

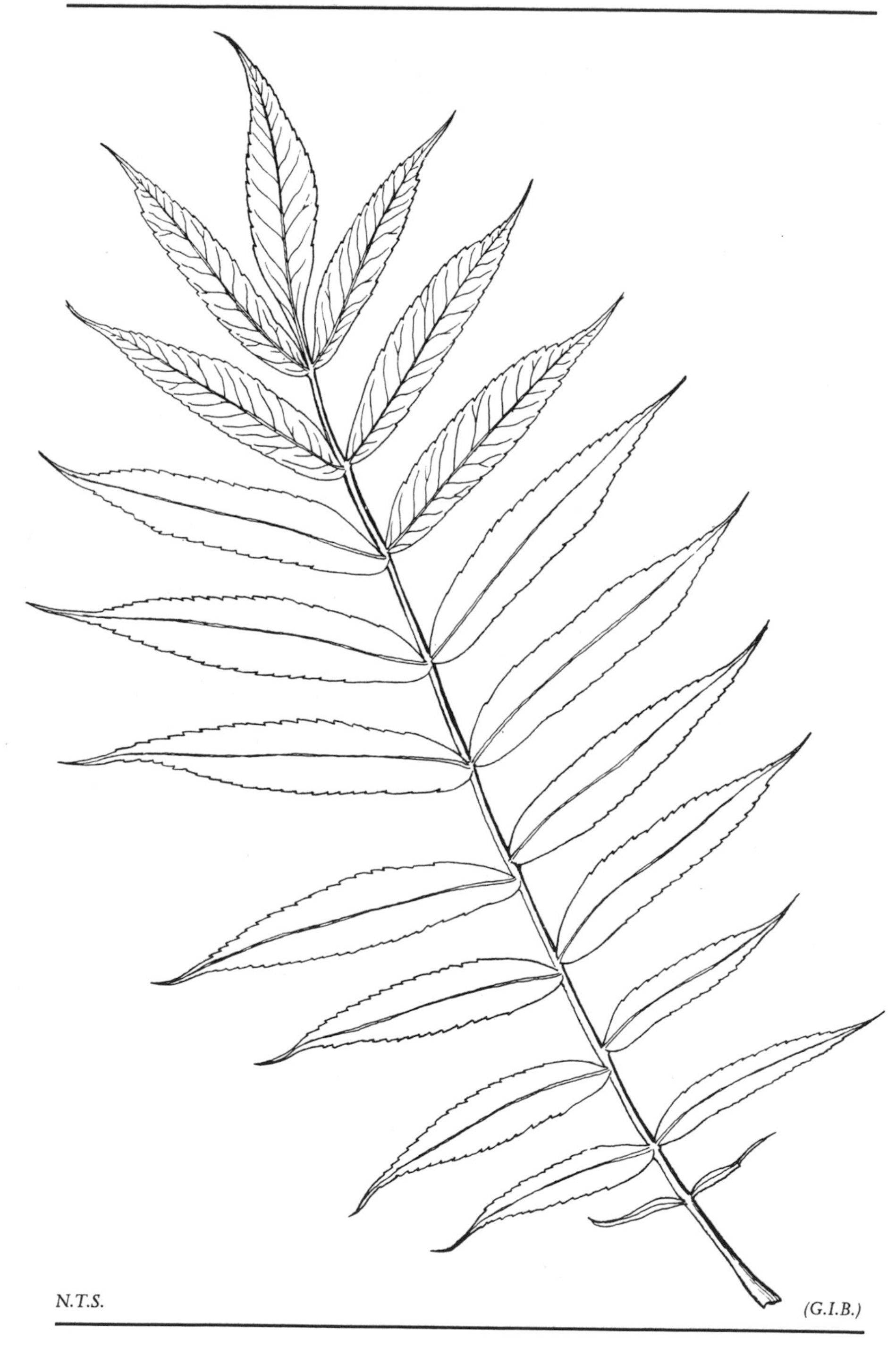

Staghorn Sumac

Staghorn Sumac

Rhus typhina L.

Range: Roadsides in New England.

Habitat: High, dry pastures and fields.

Profile: Large shrub or small tree.

Branching: Alternate.

Leaves: COMPOUND with 13 to 31, or more, sets of leaflets, long pointed, toothed, dark green and smooth above. Pale green and *woolly beneath. Midrib with small hairs.* Leaf stalk *finely hairy* 14–24" (35–60 cm) long. Leaflets have no stalks, being attached somewhat offset, directly to the leaf stem. Red in autumn.

Buds: No true terminal bud. Small, hairy surrounded by leaf scars. *Exude milky sap when broken.*

Twigs: Stout, DENSELY VELVETY HAIRY, resembling deer antlers. Pith large, brown.

Bark: Thin, brown, covered with rough dots.

Wood: Greenish satiny, soft, orange to black.

Flowers: Small white or greenish in dense hairy clusters. Turn bright red after flowering season.

Fruit: CRIMSON, DENSELY HAIRY, large cluster of berries with *long spreading red hairs.* Cone-shaped cluster, tapering at each end.

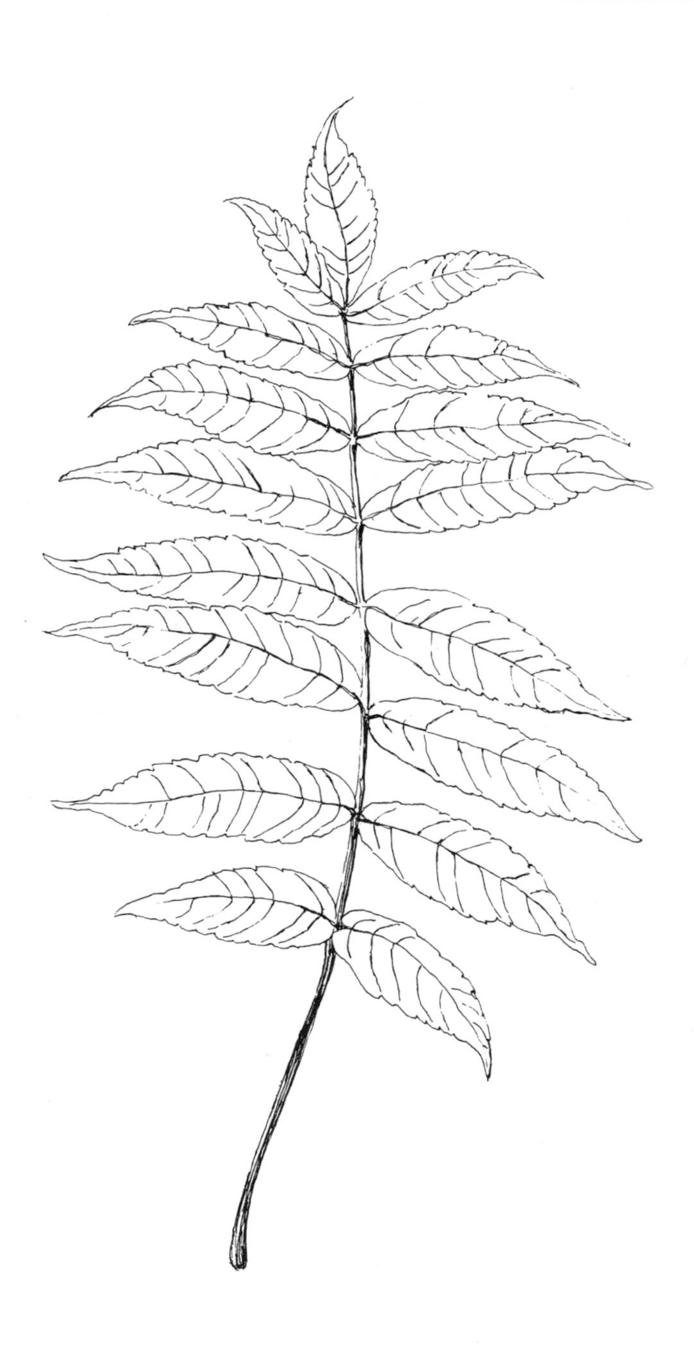

N.T.S. (P.K.)

Smooth Sumac

Smooth Sumac

Rhus glabra L.

Range: Roadsides in New England.

Habitat: High, dry pastures and fields.

Profile: Usually smaller shrub or tree.

Branching: Alternate.

Leaves: COMPOUND with 5 to 11 leaflets, usually 7, sometimes 30. Toothed, sharp pointed, bright green above much paler beneath. *Smooth above and beneath. No hairs on leaf stem or leaflets.* Often with milky juice. Lateral leaflets attached directly to the stem, the terminal one with a short stalk. Red in autumn.

Buds: No true terminal bud. Buds round, small, densely hairy, surrounded by leaf scars.

Twigs: Stout, more or *less 3-sided, smooth. Whitened by a bloom.* Pith large.

Bark: Olive to pale lavender brown.

Wood: Greenish satiny, soft, orange to black.

Flowers: More open fruiting clusters. Greenish-yellow flowers.

Fruit: Red berries, covered with sticky hairs, appearing more green, arranged in more open terminal clusters.

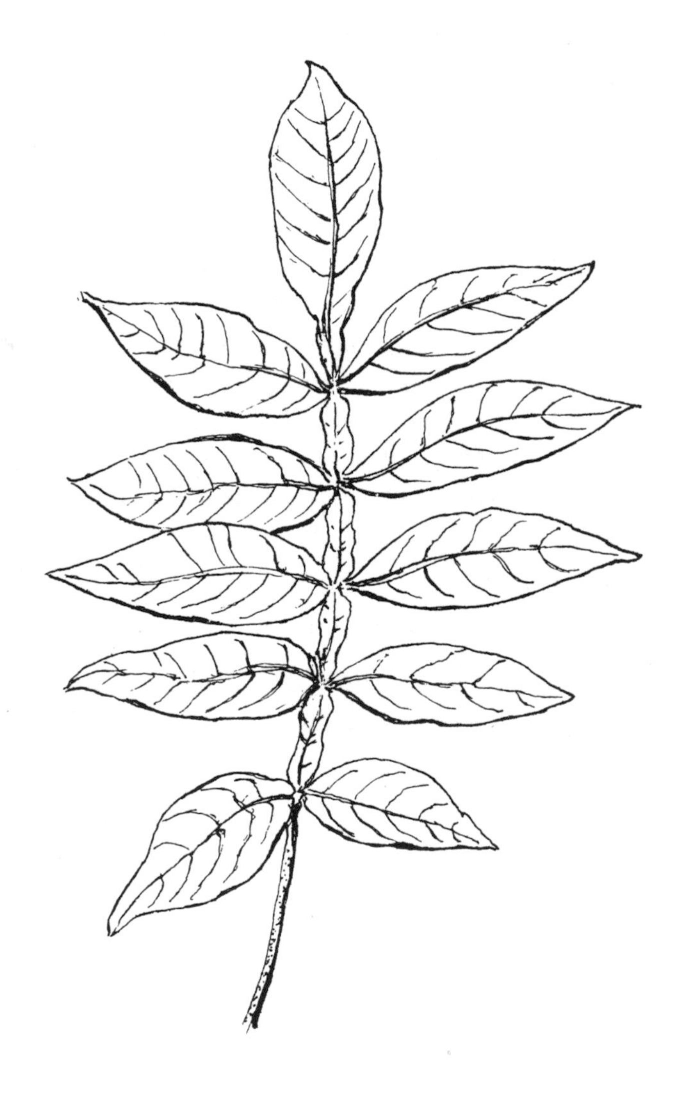

(P.K.)

Dwarf Sumac

Dwarf Sumac

Rhus copallina L.

Range: Maine to Florida and west.

Habitat: Dry, rocky or sandy soil.

Profile: Shrub up to 30' (10 m).

Branching: Alternate.

Leaves: Compound with 11–23 leaflets mostly less than 1' (20 cm) long. Leaflets pointed on both ends, smooth stalks *with winged leafy extensions.* Purple red in fall.

Buds: Terminal bud lacking. Lateral buds small, rounded, tawny to rusty.

Twigs: Stout, light reddish brown to gray sap watery.

Bark: Light brown, smooth at first, later becoming roughened and peeling off.

Flowers: Greenish white to yellowish at end of branches in erect pyramids.

Fruit: Red berries 1/8" (3 mm), thin, *fleshy, covered with red hairs, and drooping* in terminal clusters.

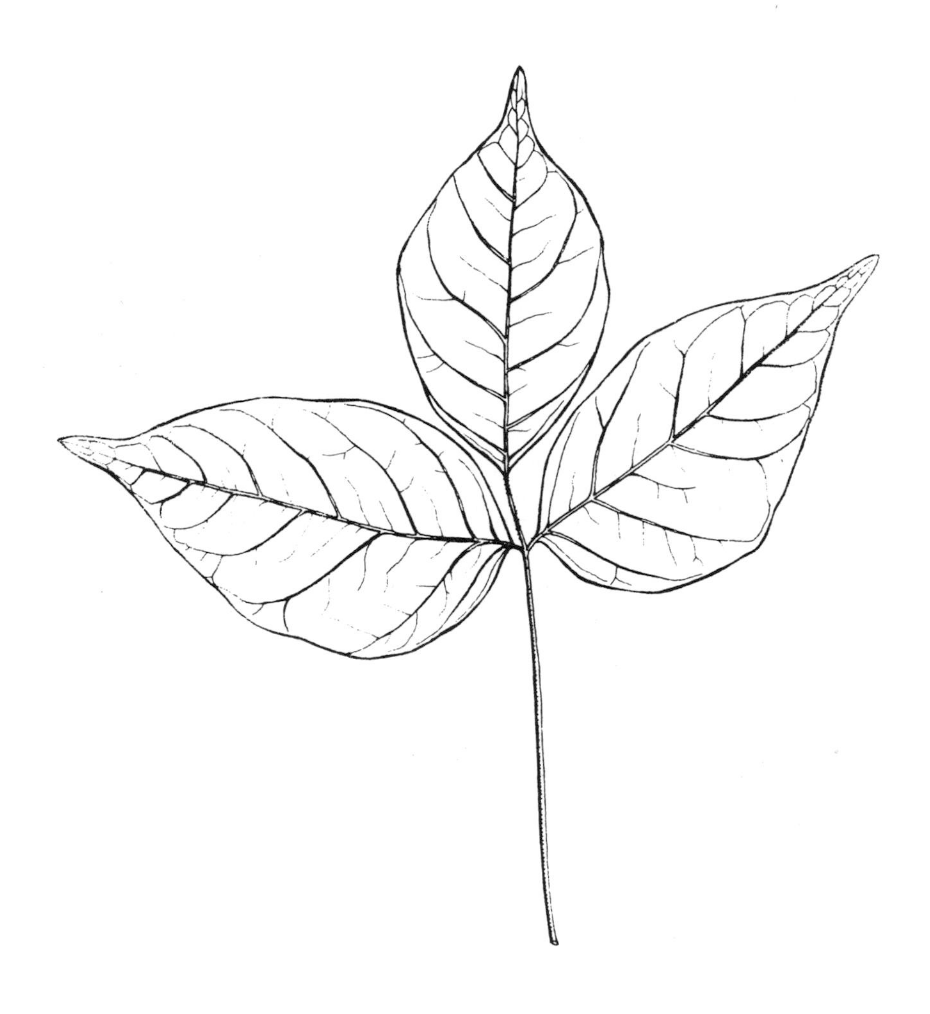

N.T.S. (G.I.B.)

Poison Ivy

Poison Ivy

Rhus radicans L.

Range: Southern New England, becoming rare northward.

Habitat: Moist, partly shaded loamy soils.

Profile: Creeping vine, sending up short, erect shoots.

Branching: Alternate.

Leaves: Compound palmate, with *3 leaflets.* Leaflets stalked entire or coarsely toothed. Ovate, *variable, often scalloped or lobed.* Terminal leaflet longer stalked than laterals. Oily looking. Red in autumn.

Buds: Hairy stalks with no scales.

Twigs: *Hairy, brownish,* with aerial rootlets.

Bark: Hairy, brown. Large stems ascending, trees may have *very long brown hairs.*

Flowers: Yellowish green, ascending.

Fruit: *Erect,* gray-white waxy berries, 1/4" (5–6 cm).

Remarks: Poisonous to touch.

(P.K.)

Poison Sumac

Poison Sumac

Rhus vernix L.

Range: Southern New England, mostly in swamps, *rare in New Hampshire.*

Profile: Shrub up to 6–20' (2–7 m).

Branching: Alternate.

Leaves: Compound, long leaflets not long-pointed, 7–13 leaflets. Obliquely ascending pointed leaflets. Smooth on margin, not silvery beneath. Leaf stalks have a clear juice turning black on drying. Red in autumn.

Buds: True terminal bud, large, conical.

Twigs: Smooth with dark dots (lenticels). Stout, yellow-brown.

Bark: *Smooth,* gray, speckled with dark dots.

Flowers: Greenish white, pendulous. Male and female on separate plants.

Fruit: *Drooping* white berries. VERY POISONOUS.

Remarks: POISONOUS TO TOUCH.

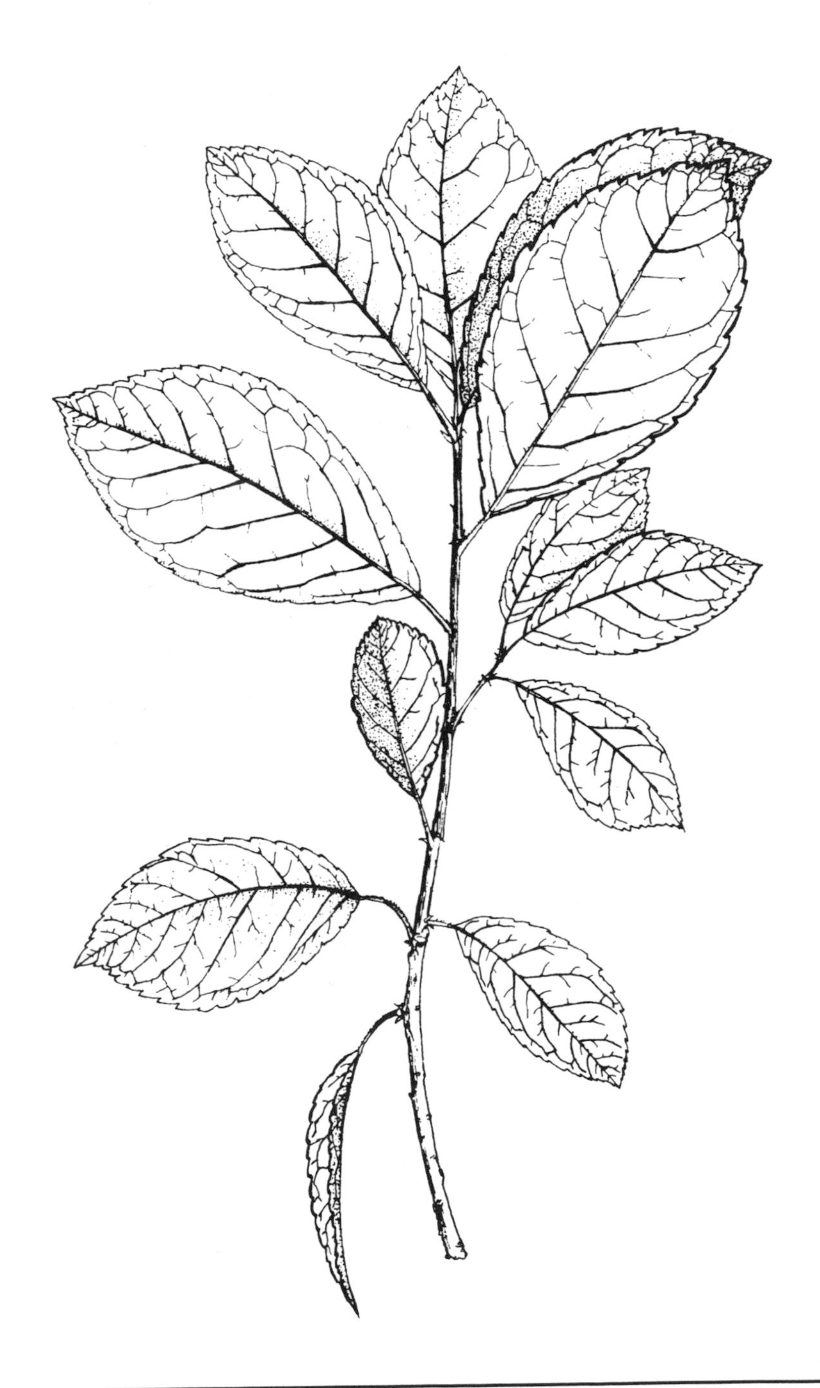

(G.I.B.)

Black Alder

Black Alder

(Winterberry)

Ilex verticillata (L.) Gray

Range: Newfoundland to Lake states, south to Georgia.

Habitat: Swamps and pond margins,

Profile: Shrub 3–15' (1–4 m) high, rarely tree-like.

Branching: Alternate.

Leading Shoot: Straight.

Leaves: Thick, dull, oval, opaque to light with teeth. Veins prominent, depressed. Veins downy beneath, coarse rounded teeth. Broadest at or above the middle. Pointed at both ends.

Buds: Very small 1/16" (2 mm) long, blunt at tip, diverging from the twig. Crowded, upper buds larger. *Terminal bud present.*

Twigs: Slender, *dark purple-gray* with many small dots (lenticels).

Bark: Gray, smooth.

Flowers: Greenish or yellowish white. Male and female on separate plants, 6 to 8 petals, in clusters seated directly on the stem, 1/4" in diameter (.5 cm) of very short stalks.

Fruit: Bright red, berry-like with 3–5 bony nuts borne singly or in clusters of 2–3" angle of branch. Persist into early winter.

Mountain Holly

Mountain Holly

(Catberry)

Nemopanthus mucronata (L.) Trel.

Range: Southern Canada, New England and southward up to 3000 ft. on New England mountains.

Habitat: Swamps and damp woods.

Profile: Large shrub.

Branching: Alternate.

Leading Shoot: Curving, twisted.

Leaves: Oblong, narrow ovate, 1–2" (2–5 cm) long, rounded at base, *finely toothed with a stem 1/2" (5–12) long,* gray blue, sometimes reddish. Thin, slightly paler beneath. Crowded, yellow in autumn.

Buds: *Terminal bud present.* Small, conical, blunt, smooth, purplish brown, 2 visible bud scales.

Twigs: Slender, smooth, greenish purple, later becoming gray, giving the bush a prominent gray appearance in winter. There are short lateral spurs.

Bark: Light gray, smooth.

Flowers: Small, greenish-white, singly in forks of the leaves in May. Sexes on different bushes. Flowers on small thread-like stalks 1/2" (1 cm) long or more.

Fruit: Berry ripening in August, but not persisting late in fall. *Red berries* on slender stalks developing in July, borne singly.

N.T.S. (G.I.B.)

Sugar Maple

Sugar Maple

(Rock Maple, Hard Maple)

Acer saccharum Marsh.

Range: Eastern U.S. and Canada.

Habitat: Moist, rich well-drained soils.

Profile: Large tree with rounded, spreading crown.

Branching: Mostly opposite.

Leaves: Large, lobed, with three main branches containing lesser rounded lobes. Flat, as wide as long, dark green, *pale beneath.* 5-lobed, *rounded between lobes, no sharp teeth.*

Buds: Sharp-pointed, scaly, 1/4" (.6 cm), reddish *brown,* often hairy, opposite on stem.

Twigs: Slender, shiny, reddish brown to *light chestnut* or gray. Current years growth *brown,* older twigs *gray.*

Bole: Straight, much-branched in open grown trees.

Bark: Light gray, GRANULAR, even on young trees, older trees gray, *hard,* deeply furrowed, *not scaly.*

Flowers: *Yellow-green without petals.* Appear in May *with the leaves.* Male and female *on same tree.*

Fruit: A double winged *brown* samara, (3 cm) long. Wings 1" (2.5 cm) long. The two wings diverge 120 degrees or less. *Ripens in fall.* Fruit stalks persist in winter. Seeds germinate during the winter or following spring.

Remarks: Hard maple group, leaves with rounded lobes.

N.T.S. (G.I.B.)

Black Maple

Black Maple

(Black Sugar Maple)

Acer nigrum Michx f.

Range: Vermont, westward and southward. Rare in New Hampshire.

Habitat: Moist, rich well-drained soils.

Profile: Same as sugar maple.

Branching: Opposite.

Leaves: Deeper green than sugar maple, *weaker,* margins and lower lobes *drooping,* yellow-green beneath. Lower *surfaces and stems hairy.* Mostly *3-lobed,* rarely 5-lobed. *Lobes narrow.* Thicker than sugar maple. Lobes much broader and shorter, with few undulations, often no teeth.

Buds: Large, brown, hairy.

Twigs: Stout, warty, streaked. Hairy when young, later smooth, orange or yellow all the year.

Bole: Similar to sugar maple.

Bark: *Darker* than sugar maple, furrows narrower and more shallow than sugar maple. Black on old trees.

Flowers: Similar to sugar maple.

Fruit: Similar to sugar maple, but *more divergent.*

Remarks: Often considered a variety of sugar maple, rather than a separate species.

N.T.S. (G.I.B.)

Norway Maple

Norway Maple

Acer platanoides L.

Range: Western Europe.

Habitat: Widely planted as a city shade tree.

Profile: Round-crowned, spreading branches.

Branching: Opposite.

Leaves: Very *dark green to bronze,* smooth, 5-lobes with a few large teeth (rarely 7). *Shiny beneath,* points of lobes *sharp-pointed,* more so than sugar maple. Leaves usually larger than sugar maple. Stem with *milky juice.*

Buds: *Large reddish,* sometimes mixed with green. Terminal bud larger than laterals. Scales hairy, dark rusty.

Twigs: Coarse, smooth, shining, brown to yellowish brown. Streaked with fine longitudinal cracks in the bark.

Bole: Straight, much branched.

Bark: Smooth, closely furrowed not scaly. Narrow ridges run together to form diamond-shaped spaces, somewhat resembling white ash, but finer.

Flowers: Greenish-yellow appearing with or before leaves. Petals conspicuous. Flowers stalked.

Fruit: Over 2" (5 cm) long, seed portion flattened, wings DIVERGING IN ALMOST STRAIGHT LINE.

Remarks: These descriptions demonstrate the differences between the hard maple and soft maple groups. They are placed together for easy comparison. Other species of maple are similar.

N.T.S. *(G.I.B.)*

Red Maple

Red Maple

(Soft Maple, Swamp Maple)

Acer rubrum L.

Range: Eastern U.S. and Canada.

Habitat: Swampy sites and moderately moist, sandy loam, and even on rocky uplands.

Profile: Medium-sized tree with ascending branches

Branching: Opposite.

Leaves: 5-lobed with SHARP ANGLES BETWEEN LOBES, and *with sharp irregular teeth.* Dark green above, gray green beneath.

Buds: Sharp, *dark red,* with prominent overlapping scales.

Twigs: Current years growth *red.* Older twigs light gray, smooth.

Bole: Straight, slender, small taper in forest-grown trees.

Bark: Gray, *smooth, often satiny* on young trees. Scaly, darker on older trees becoming soft, rough, separating and forming long scales, often up-curling.

Flowers: *Red* or yellow with *petals* appearing *long before the leaves.* Male and female on *separate trees.*

Fruit: A double-winged, V-shaped *red* samara. Wings 3/4" (2 cm) long. Matures *in late spring*, falls and *germinates at once.*

Remarks: Soft Maple group, *leaves with sharp angles in lobes.* These descriptions demonstrate the differences between the hard maple and soft maple groups. They are placed together for easy comparison. Other species of maple are similar.

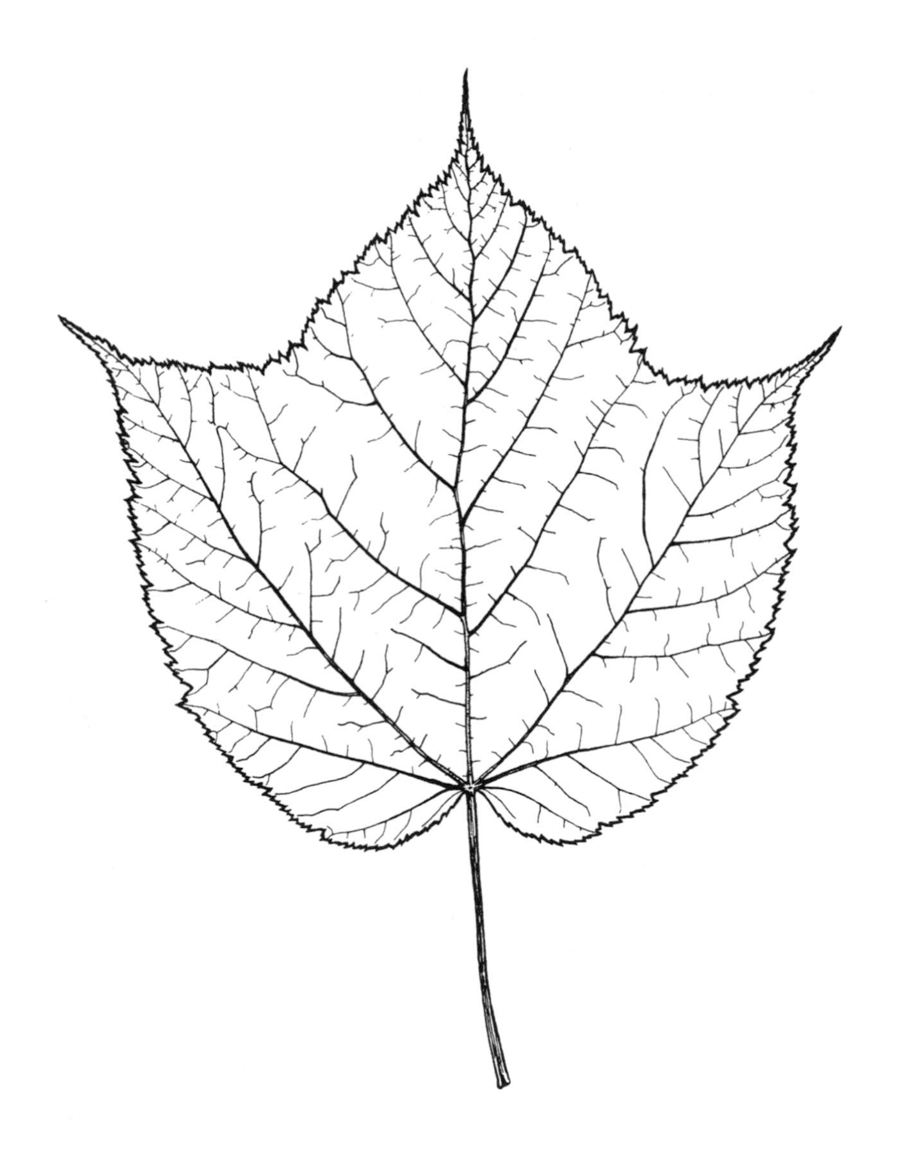

N.T.S. *(G.I.B.)*

Striped Maple

Striped Maple

(Moosewood, Whistlewood)

Acer pensylvanicum L.

Range: Eastern U.S. and Canada.

Habitat: Upland woods.

Profile: Small, slender tree

Branching: Opposite.

Leading Shoot: Erect.

Leaves: Very *large,* 3 main lobes, pale green to yellowish green margins with small *regularly* spaced teeth. Tips to lobes long pointed. Broadest above middle, *yellow* in autumn.

Buds: Terminal bud bright red, large and stalked. Large angular buds.

Twigs: New shoots reddish, with *white streaks* gradually forming. A few small white dots BECOMING GREEN WITH WHITE STREAKS.

Bole: Small, erect, rarely in clumps.

Bark: Smooth, *greenish,* STREAKED WITH BLACK AND WHITE, mostly *white streaks.*

Flowers: Greenish yellow, 3–4" (8–10 cm) broad unfolding after the leaves, long drooping clusters. Sexes on different trees or on the same tree.

Fruit: Widely divergent maturing in July. *No ribs over the seed.*

N.T.S. (G.I.B.)

Mountain Maple

Mountain Maple

Acer spicatum Lam.

Range: Northern U.S. and Canada, higher elevations in southern part of range.

Habitat: Cool woods

Profile: Tall shrub or bushy, small tree.

Branching: Opposite.

Leading Shoot: Erect.

Leaves: Thin, *herbaceous* in texture, 3-lobed, teeth with *rounded margins.* Veins prominent; downy beneath. *Scalloped shape between teeth.* Brilliant *red* in autumn.

Buds: Small, short stalks, shiny 2–4 small scales, inner pair hairy.

Twigs: *Gray, with fine hairs,* often streaked with green toward the base end with terminal bud

Bole: Usually in dense clumps like a shrub.

Bark: Thin, smooth, light grayish brown, *often mottled.*

Flowers: Greenish yellow unfolding after the leaves, small in *stiffly erect* clusters up to 3" (7 cm) long.

Fruit: Red, broad *ribbed* maturing *seed* in July.

N.T.S. (G.I.B.)

Silver Maple

Silver Maple

(White, Soft or River Maple)

Acer saccharinum L.

Range: Eastern Canada to Minnesota and southward.

Habitat: River banks and bottomlands.

Profile: Slender, drooping large tree, branching low and spreading into a broad crown.

Branching: Opposite.

Leading Shoot: Drooping. Grow downward, then curve up.

Leaves: Opposite, 5" (12 cm) in diameter. *Deeply cleft,* 5-lobed, serrate margin sharply toothed. Green above, *silver-white below.* Very silvery under water. Downy beneath when young. Pale yellow in autumn.

Buds: Terminal bud small, blunt, red.

Twigs: V-shaped, point upward at tip. Drooping. Orange-brown to red. *Rank, disagreeable odor when bruised.*

Bole: Often bent, wavy or bending.

Bark: *Scaly.* Smooth in young trees breaking into long thin scaly plates, curving away at ends. Flaky.

Flowers: Red or greenish yellow. *No petals. Appearing long before the leaves.* Some trees have only male flowers.

N.T.S. (G.I.B.)

Box Elder

Box Elder

(Ash-Leaved Maple)

Acer negundo L.

Range: Eastern U.S. to southern Canada. In New Hampshire only along Connecticut River and western areas.

Habitat: River banks.

Profile: Erect, much branched small tree.

Branching: Opposite.

Leading Shoot: Erect

Leaves: *Compound palm-like leaves* with 3–5 veiny *leaflets, on short stalks. Very variable* in shape, tip of leaflets, sharp margin with coarse teeth. Light green above, pale green below, hairy along veins below. Stalk swollen at base.

Buds: Short-stalked, reddish, usually woolly or downy, whitish, densely hairy.

Twigs: Stout, green to purplish green, brightly colored, often red. Covered with whitish bloom the first year, which readily rubs off.

Bole: Usually poor form.

Bark: Thin, light brown, with narrow ridges suggestive of Norway maple.

Flowers: Male and female flowers on separate trees, appearing slightly before leaves. Greenish in stalked clusters. Female flowers in drooping clusters, 6" long.

N.T.S. (G.I.B.)

Horse Chestnut

Horse Chestnut

Aesculus hippocastanum L.

Range: Introduced from Europe and Asia.

Habitat: Introduced as an ornamental tree.

Profile: Large tree, bole continuing to the top, or dividing.

Branching: Opposite.

Leaves: Compound, palm-like, 5–7 leaflets diverging from a center. Leaflets straight-veined.

Buds: Large, nearly black, varnished with sticky gum, opposite. Terminal bud present.

Twigs: Stout, red-yellow to brown. Slightly downy.

Branches: Lower branches drooping with upturned tips.

Bark: Dull brown, slightly fissured.

Flowers: Erect clusters of white flowers like columns, 4–5 petals, red-marked at base.

Fruit: A large thick, fleshy, weak-spined burr containing a large shiny brown nut. Bitter tasting.

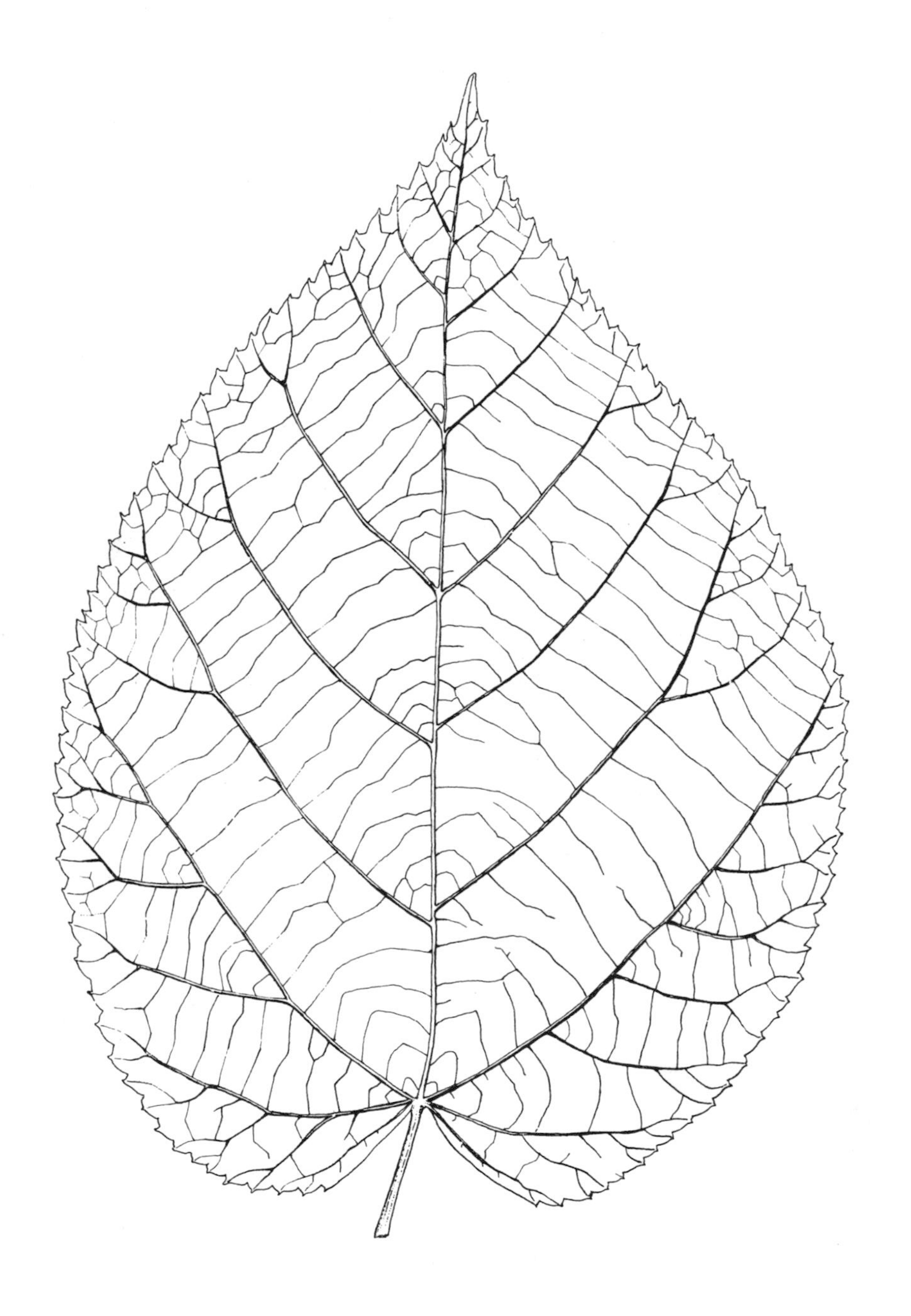

N.T.S. (G.I.B.)

Basswood

Basswood

(American Linden)

Tilia americana L.

Range: Eastern and central U.S. and southern Canada.

Habitat: Rich woods.

Profile: Large tree.

Branching: Alternate.

Leading Shoot: Straight.

Leaves: Large, up to 8" (20 cm), sharply toothed. *Unequally heart-shaped at base* under surface *with tufts of hair* in corners of *lateral veins,* but wanting in those at base of leaf.

Buds: Carmine or red-brown, 1/2" (1 cm) long, no terminal bud. *Lopsided.*

Twigs: Alternate, greenish-gray, zigzag, smooth, stout, smooth or downy. Pith brownish buff.

Branches: Straight.

Bole: Straight, often clear of branches, USUALLY MULTIPLE STEMMED.

Bark: Smooth, gray, ridged when old.

Flowers: 1/2" (1 cm) across, 5 yellow petals. July, cream-colored, honey bearing, fragrant.

Fruit: Stalked cluster of small pale green nuts, *attached to a leafy wing.*

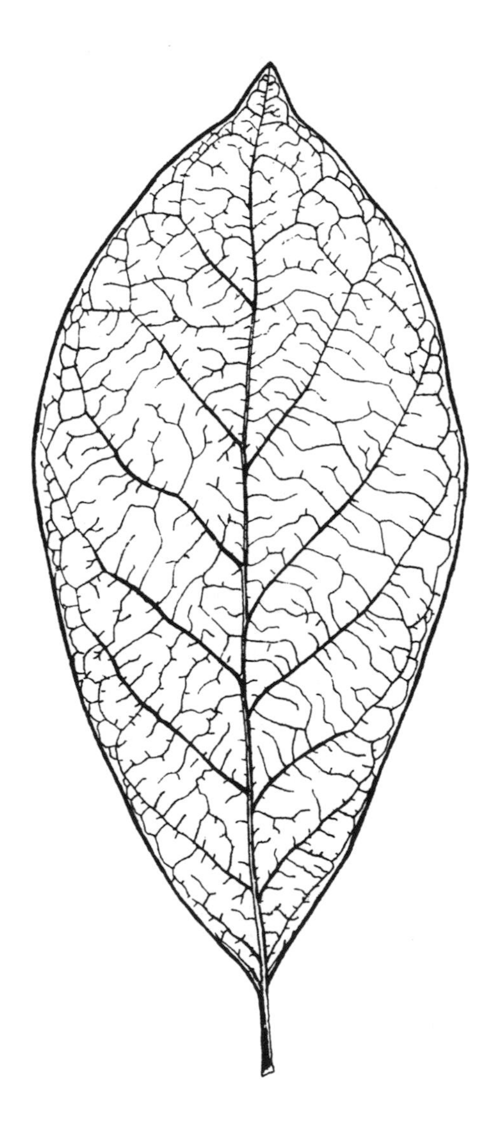

(G.I.B.)

Black Gum

Black Gum

(Sour Gum, Pepperidge, Black Tupelo, etc.)

Nyssa sylvatica Marsh.

Range: Southern New England, New York, and south and westward.

Habitat: Damp and swampy ground, southern highlands.

Profile: Medium-sized tree.

Branching: Alternate.

Leading Shoot: Crooked.

Leaves: About 4" (10 cm) long, oval, *no teeth, leathery,* shiny above. *Often single leaves turn red and fall in July,* all scarlet in autumn. Broadest above middle. Abrupt short tip.

Buds: Smooth, dark, red-brown.

Twigs: Slender, short spur shoots with dwarfed branchlets. Pith with faint bars of darker tissue.

Branches: Stiff, straight, horizontal branches, making a dense crown.

Bole: Straight, clear of branches, one stem.

Bark: Light brown to gray, becoming *deeply fissured when old,* BREAKING INTO DENSE, HARD OBLONG BLOCKS.

Flowers: Inconspicuous, greenish, several flowers per stalk. Different sexes on different trees. Male flowers small, female larger, smaller clusters.

Fruit: Blue berry, single large stone, sour, mature in July.

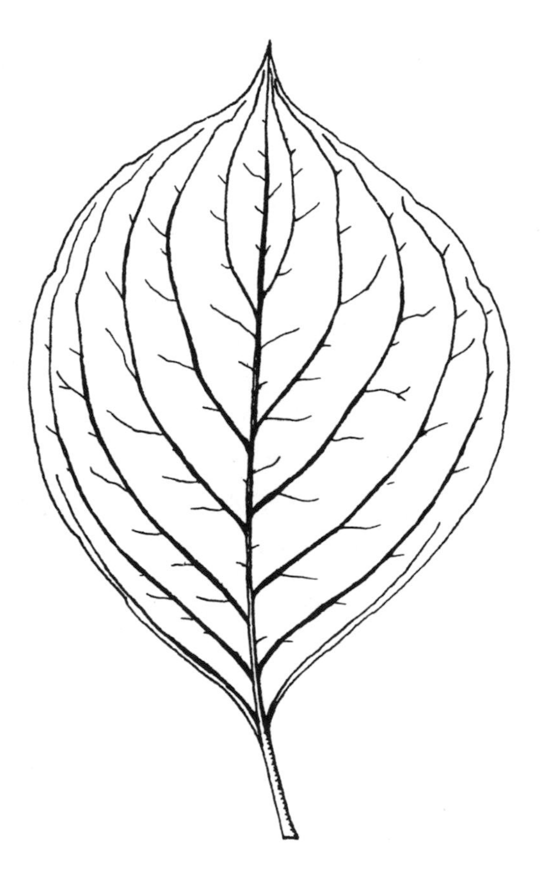

(G.I.B.)

Flowering Dogwood

Flowering Dogwood

Cornus florida L.

Range: Southern Maine, New Hampshire, south to Florida.

Habitat: Acid woods.

Profile: Large shrub to small tree 10–40' (3–12 m).

Branching: Opposite.

Leaves: Smooth, elliptic or wedge-shaped with *5–6 pairs of veins* 2–5" (5–12 cm) long. *Tip tapering,* rounded at base. No hairs beneath.

Buds: *Flower buds stalked,* resemble miniature urns. Side buds hidden.

Twigs: Red-green to *dark purple,* angled, glaucous bloom. Appear mealy, pith white, gritty, granular.

Bole: Straight, short, clear length.

Bark: Dark gray, deeply checkered, broken into small *blocks with an alligator hide-like pattern.*

Flowers: Small, clustered. 4 (or rarely more) *large white to pink bracts that look like petals.*

Fruit: Red (or rarely yellow) berries, 3/8" (1.5 cm) long in small *clustered heads.*

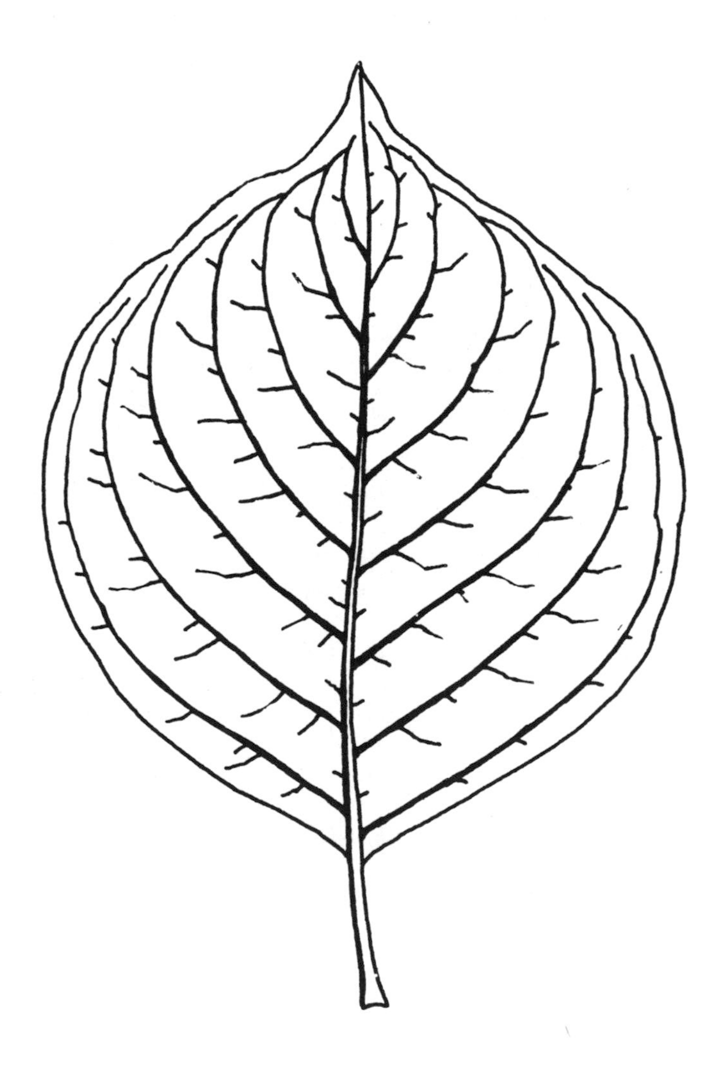

(G.I.B.)

Round-Leaved Dogwood

Round-Leaved Dogwood

Cornus rugosa lam.

Range: Quebec to Virginia.

Habitat: Dry, rocky woods.

Profile: Coarse shrub, 3–10' (1–3 m).

Branching: Opposite.

Leaves: Broad, ovate with an *abrupt sharp* point, rounded at base, 7–9 *pairs* of veins, *woolly beneath,* wooly to the touch.

Buds: Short-stalked, oblong, pointed, hairy.

Twigs: Young branches greenish, *blotched with purple.* Large white pith.

Bole: Short, shrub-like.

Bark: Not different from twigs.

Flowers: White clusters, small in flat-topped clusters, 3" (8 cm) across.

Fruit: Round, lead-colored, about 1/4" (.5 cm) in red-stemmed clusters, 2–3" (5–8 cm) in diameter.

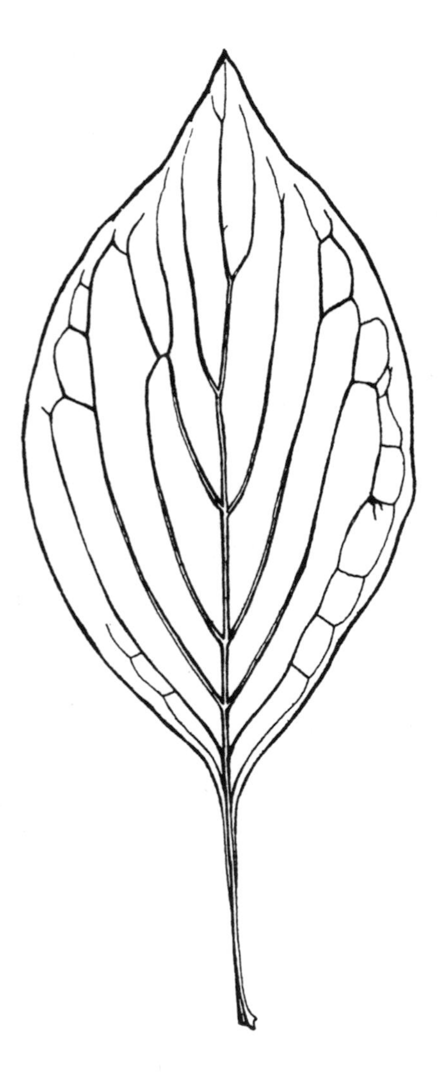

(G.I.B.)

Alternate-Leaved Dogwood

Alternate-Leaved Dogwood

(Cornel, Pagoda Dogwood)

Cornus alternifolia

Range: Eastern U.S. and Canada.

Habitat: Deep woods.

Profile: Small, erect shrub.

Branching Alternate

Leading Shoot: Straight, short.

Leaves: Crowded at the tip of the branch. *Prominently veined strongly curved to parallel the edges.* No teeth. Smooth and pale beneath.

Buds: Narrow, dark red.

Twigs: Red-brown with diamond-shaped gray spots. Somewhat zigzag. Pith white.

Flowers: Hemispheric. Petals variable, 1–2" (8–50 mm) on same stem.

Fruit: Blue. Stone deeply pitted.

N.T.S. (G.I.B.)

Red Osier Dogwood

Red Osier Dogwood

Cornus stolonifera

Range: Northern New England and Canada, South-Central states, westward.

Habitat: Stream banks.

Profile: Low spreading shrub.

Branching: Opposite.

Leading Shoot: Long, straight.

Leaves: About 4" (10 cm) long, elongated, oval, with prominent veins. Whitish beneath, hairy on both surfaces.

Buds: Pointed, hairy.

Twigs: Recent wood, DEEP RED. Pith white.

Branches: Spreading.

Flowers: Small, white, in flat-topped clusters, 2" (5 cm) broad.

Fruit: White or lead-colored.

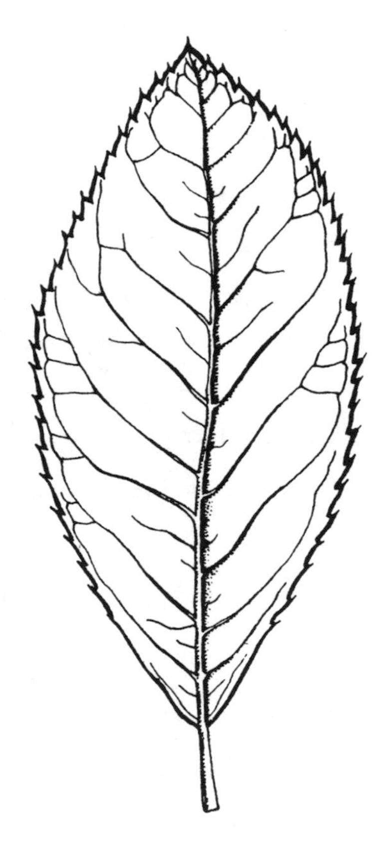

(G.I.B.)

Sweet Pepperbush

Sweet Pepperbush

Clethra alnifolia L.

Range: Southern Maine and New York, south to Florida.

Habitat: Swamps and low sandy woods.

Profile: Large shrub.

Branching: Alternate.

Leaves: Small, 1 1/2–3" (3–7 cm), broadest above middle, sharply toothed except at base. Straight-veined; both sides smooth and green, blunt-pointed.

Buds: Small and obscure except terminal, which is downy, large and pointed.

Twigs: Slender, round and slightly downy on new twigs. Pith large and white.

Bark: Dark grayish or blackish.

Flowers: VERY FRAGRANT. Small white, sometimes pinkish, 1/2" (1 cm) across in long spike clusters. July–September.

Fruit: Depressed round balls, 1/8" (.5 cm) tipped with a threadlike spike remaining from the flower. The dry pod splits into 5 parts.

(P.K.)

Pale Laurel or Bog Laurel

Pale Laurel or Bog Laurel

Kalmia polifolia Wang.

Range: Labrador to Alaska, south to New Jersey and Pennsylvania.

Habitat: Peat bogs.

Profile: Slender, small shrubs up to 2' (.6m).

Branching: *Opposite.*

Leaves: Thick, stiff, leathery, evergreen, *opposite,* (rarely in threes), *no leaf stalks. Inrolled margins,* shiny green above, *white* beneath, midrib prominent beneath. Tips blunt.

Buds: Inconspicuous.

Twigs: *2-edged with direction changed at each node.* Smooth, yellow-brown.

Flowers: Bluish-pink or crimson, with papery overlapping lobes, 1/2–1" (1–2 cm) broad. Clustered at ends of stems.

Fruit: Erect, globe-like, falling early. A many seeded pod 1/8".

(G.I.B.)

Bog Rosemary

Bog Rosemary

Andromeda glaucophylla Link.

Range: Greenland to Labrador, south to Pennsylvania.

Habitat: Peat bogs.

Profile: Low shrub with creeping base up to 2' (.6 m).

Branching: Alternate.

Leaves: Thick, leathery evergreen, 3/4–2" (2–5 cm) long, 1/4" (.6 cm) wide. *Alternate. Very narrow, inrolled margins.* Pale, blue-green above, very *white beneath* with very fine hairs. TIPS SHARP.

Buds: Bud scales whitened.

Twigs: *Round, much whitened.*

Flowers: Small, pink or whitish, urn or turban-shaped in terminal clusters on thick, curved stalks. Twice the length of the flower.

Fruit: Small, round whitish pods covered with white bloom, 5-seeded.

(G.I.B.)

Leatherleaf or *Cassandra*

Leatherleaf or Cassandra

Chamaedaphne calyculata (L.) Moench, *var angustifolia* (Ait.) Rehd.

Range: Eastern Canada to Alaska, south to Georgia.

Habitat: Peat bogs and pond margins, confined to wet areas.

Profile: Low, much-branched shrub up to 3' (1 m).

Branching: Alternate.

Leaves: Evergreen or nearly so, oblong, 1–2" (2.5–5 cm) long, narrowed at base, slightly inrolled, no teeth. Upper *surface covered with scaly dots* giving a roughened appearance. *Rust-colored beneath.* Reddish in winter.

Twigs: New twigs covered with minute scales, older stems dark copper colored and smooth.

Flowers: Small, white, tinged with purple, urn-shaped, *hanging in lines of 10–30 blossoms under a leafy stem.* April.

Fruit: Many seeded capsules, *persistent in winter,* containing flattened wingless seeds.

(P.K.)

Labrador Tea

Labrador Tea

Ledum groenlandicum Oeder

Range: Eastern Canada to Alaska.

Habitat: Bogs, alpine tundra.

Profile: Low shrub up to 2' (.6 m).

Branching: Alternate.

Leaves: Oblong, narrow, blunt, rough, evergreen with margin, STRONGLY INROLLED with a dense BROWN, RUSTY WOOL BENEATH. *Fragrant when bruised.*

Twigs: Recent shoots with rusty hairs, older stems very dark-reddish brown.

Flowers: Erect white with 5 spreading petals on long stalks forming flat clusters at ends of branches.

Fruit: Slender cylinders 1/4" (.6 cm) long, nodding, with 5-celled opening at base.

Highbush Blueberry

Highbush Blueberry

Vaccinium corymbosum L.

Range: Southern Canada west to Wisconsin and southward.

Habitat: Swamps, woods, dry pastures.

Profile: Large shrub.

Branching: Alternate.

Leading Shoot: Curving, compact, open clumps.

Leaves: Oval, green on both sides, *no teeth on margin.* Lower side slightly woolly on the nerves. Pointed at both ends. *Deep red in autumn.*

Buds: *True terminal bud lacking.* Flower buds red, plump, pointed; the scales with a spine-like point.

Twigs: Yellowish green, becoming pale red. Some are reddish brown, on other bushes yellow. *Covered with minute warty dots, zigzag or angled.*

Bark: Rough brown.

Flowers: Urn-shaped, white or pink, 1/3" (7 mm) long, May or June *when leaves are only half grown.*

Fruit: *Blue-black berry, sweet,* in clusters.

(G.I.B.)

Sheep Laurel

Sheep Laurel

(Lambkill)

Kalmia angustifolia L.

Range: Labrador to Manitoba, south to mountains of Georgia.

Habitat: Wet or dry acid soil, barrens.

Profile: *Slender* shrub up to 3' (1 m) high.

Branching: Mostly *opposite.*

Leading Shoot: Straight.

Leaves: *Evergreen.* Flat, thin, narrow, oblong on very short stems, *crowded* on *the branch opposite* in threes. Upper surface bright green, lower pale, or whitened. Both surfaces smooth.

Buds: Naked, no scales, or only 2 showing.

Twigs: *Round* in cross section. Pale brown, smooth.

Branches: Opposite, curving.

Bark: Reddish brown.

Flowers: *Deep pink,* small 1 1/2" (1 cm) across, *in clusters among the leaves.* Rose-pink or crimson.

Fruit: Small, many-seeded round pod, 1/8" (3 mm) in diameter, recurving.

(G.I.B.)

Mountain Laurel

Mountain Laurel

(Calico Bush, Spoonwood)

Kalmia latifolia L.

Range: Central and southern New England to Florida.

Habitat: Rocky slopes, sometimes swamps.

Profile: *Coarse shrub* or small tree up to 15' (4.5 m).

Branching: Mostly *alternate.*

Leading Shoot: Curving

Leaves: *Evergreen.* Flat, leathery, broad, narrow at both ends. Mostly *alternate* on the branch, sometimes opposite. In clusters at the ends of stems.

Buds: Naked.

Twigs: Smooth and green first year, later reddish brown.

Branches: Curved, alternate.

Bark: Reddish brown, thin, peeling in narrow shreds.

Flowers: *Terminal. 20 or more* in clusters up to 3" (7 cm), broad, *white* or pink, cup-shaped with 5 petals. June.

Fruit: Pod 1/4" (6 mm) in diameter, ascending.

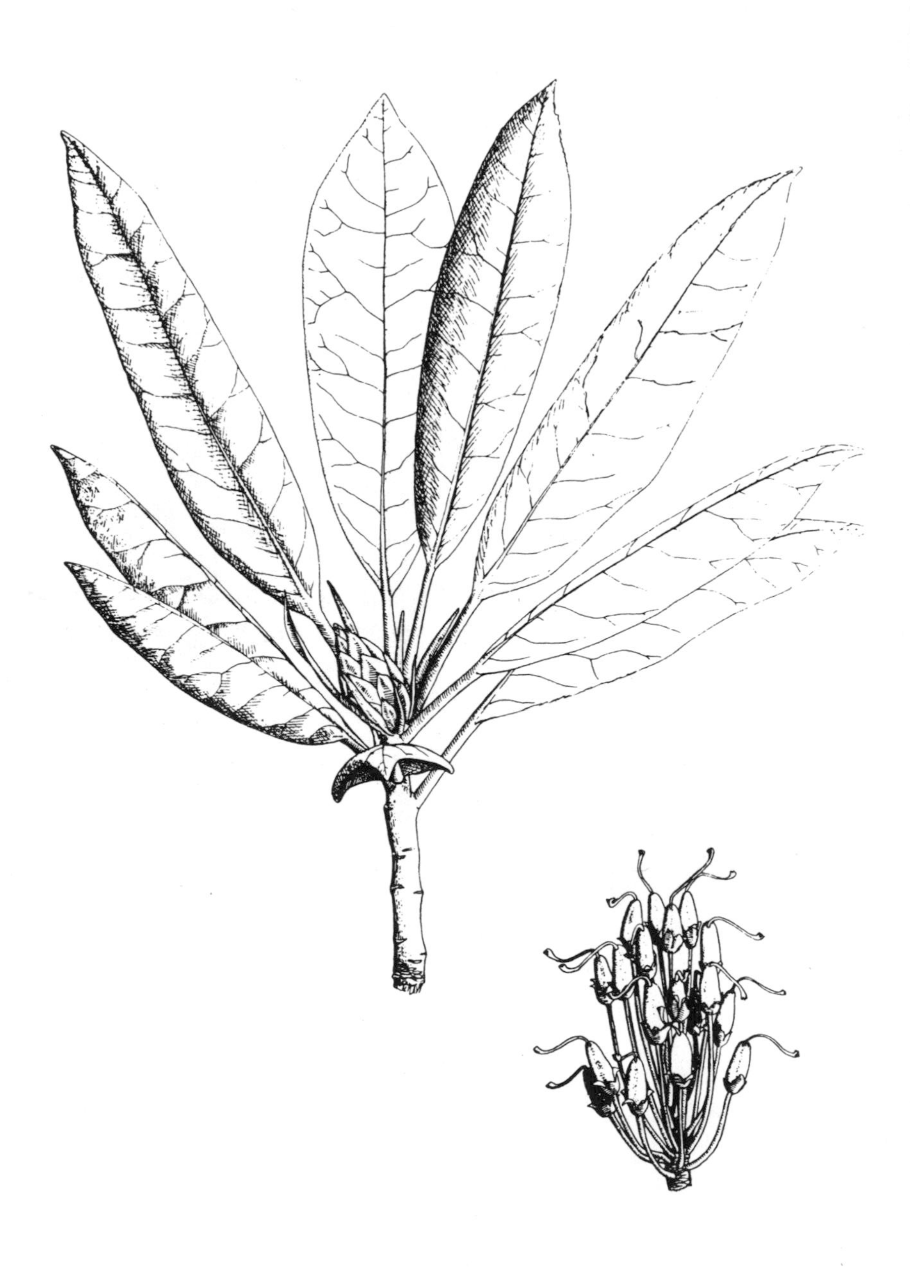

N.T.S. (G.I.B.)

Rhododendron

Rhododendron

(Great Laurel, Rosebay)

Rhododendron maximum L.

Range: New England south to Georgia, Alabama, Virginia.

Habitat: Damp woods, swamps and pond margins, higher elevations south.

Profile: Large shrub or small tree up to 30' (10 m).

Branching: Alternate.

Leaves: EVERGREEN, very thick, leathery, 5–10" (12–25 cm) long and 1–3" (2–8 cm) broad, pointed at each end, without teeth on margin; dark green above, paler and often hairy beneath. A bristle at point of leaf. Leaf stalks, stout, short and hairy. Tend to curl and droop in winter.

Buds: Small, inconspicuous. Flower buds large, scaly, pale green at ends of branches.

Twigs: Young branches hairy.

Bark: Reddish brown, peeling in shreddy scales.

Flowers: Large in numerous clusters, rose-pink to white, flower stalks sticky. Deeply cleft into 5 lobes, sprinkled with yellow or orange spots. July.

Fruit: A dry, oblong pod, sticky.

N.T.S. (G.I.B.)

Rhodora

Rhodora

Rhododendron canadense (L.) Torr.

Range: Eastern Canada, south to New York and northern New Jersey.

Habitat: Bogs, rocky summits, slopes.

Profile: A *low shrub* with straight, upright branches up to 3' (1 m), compact and bushy.

Branching: Alternate.

Leaves: *Deciduous* (not evergreen) 2" (5 cm) long, oblong, narrowed at base. Margins *slightly rolled without teeth.* Gray-green above, smooth, somewhat downy and paler beneath or whitened rusty hairs along midrib.

Buds: Light lavender brown with whitish bloom.

Twigs: Pinkish or pale yellow-red with whitish bloom, slender, smooth, slightly crooked.

Bark: Outer bark peels off leaving copper-colored, smooth inner bark.

Flowers: Appear before the leaves in April or May. Pale pink to rose-purple, *mostly odorless,* 2-lipped with 5 recurving petals in clusters at top of branch, 1" (2.5 cm) long and 1/2" (1 cm) across on short hairy stalks.

Fruit: Oval, glandular capsule, 1/2" (1 cm) long, in clusters, dividing into 5 segments. *Lopsided at base.*

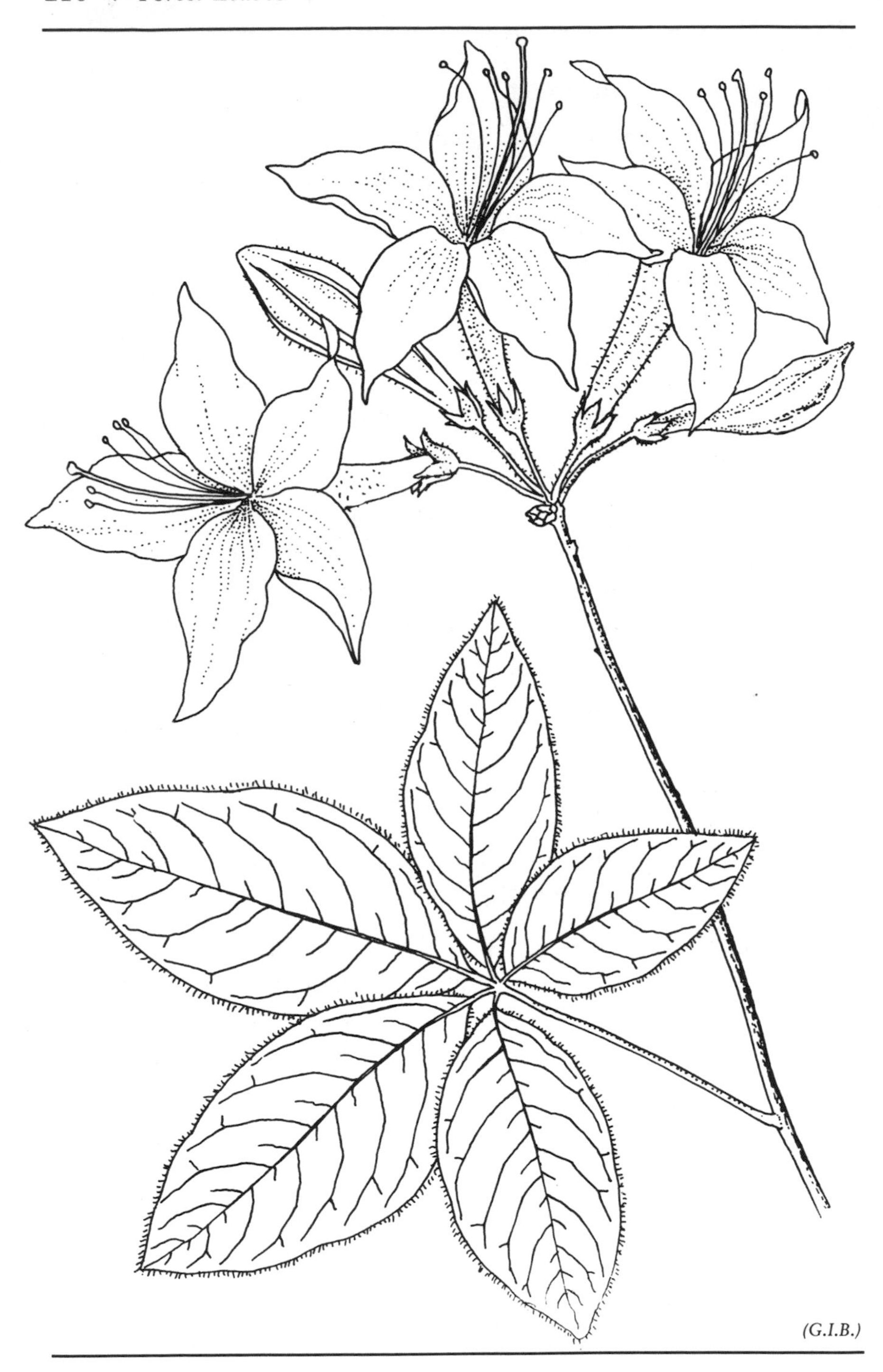

Early Azalea

Early Azalea

Rhododendron roseum (Loisel.) Redh.

Range: Southwestern Maine to Virginia, Tennessee, Missouri.

Habitat: *Dry* woods, rocky slopes.

Profile: Tall shrub up to 9' (3 m).

Branching: Alternate.

Leaves: Up to 3" (8 cm) long, over 1/2" (1 cm) wide, *(shorter than pinxter flower)*, clustered at the branch tips. *Hairy beneath,* somewhat *woolly above,* sometimes with fine teeth, not rolled on margins.

Buds: Large, gray-brown, 1/4" (6 mm) flower buds.

Twigs: Buff to ashy-brown, with slight *gray down or wool.*

Flowers: VERY FRAGRANT bright pink to white *appearing with the leaves.* Flower is lipped with 5 petals, each about 2" (5 cm) long. Clustered at the ends of stems.

Fruit: A pod 1/2" (1 cm) splitting into 5 parts, persistent in winter, leaves shorter and broader than Pinxter flower.

N.T.S. (P.K.)

Pinxter Flower, Election Pink

Pinxter Flower, Election Pink

Rhododendron nudiflorum (L.) Torr.

Range: Southern New England, southward.

Habitat: Moist woods, rocky stream banks, swamps.

Profile: Tall shrub up to 9' (3 m). *Much branched at top.*

Branching: Alternate.

Leaves: Oblong, 1½–3" (3–8 cm) long. *Green on both sides,* often *finely toothed.* Broadest above the middle. *Smooth on both sides.* A few hairs on mid ribs only.

Buds: Large flower buds at tips of branches. Buds pale greenish, tinged with red.

Twigs: Slender, smooth, buff brown to gray.

Flowers: Sweet delicate fragrance when fully grown, but mostly *odorless.* Appearing BEFORE THE LEAVES. Pale pink to white. The slender tube flares above and is longer than the lobes. Flower stems erect, hairy.

Fruit: Oblong, erect, opening down from the top. Persistent in winter.

(G.I.B.)

Maleberry

Maleberry

(Privet Andromeda, Male Blueberry)

Byonia ligustrina (L.) DC.

Range: New England south to South Carolina.

Habitat: Swamps and damp woods.

Profile: Shrub 2–6' (1–2 m) high. Much branched.

Branching: Alternate, crooked.

Leaves: 1/2" (2.5–5 cm) long, pointed both ends, *minutely toothed, or entire* with short stem. Rough, *scurfy on upper surface,* slightly hairy beneath.

Buds: *No terminal bud.* 1/8" (5 mm) pointed with a single reddish scale.

Twigs: *Yellow-brown* to ashy gray mottled with black. *Light-colored shedding bark,* flattened at joints.

Bark: Light-colored, shedding. Gray, shaggy.

Flowers: Small, globe-shaped, nodding, white, about 1/8" (3 mm) across, up to 50 in a cluster. Flowering twigs, leafless.

Fruit: Greenish gray to brown, small, globe-like capsules with 5 cells. No larger than flower. *Crowded in groups on the stem, remaining all winter.* Dry, brown pods, resembling blueberries.

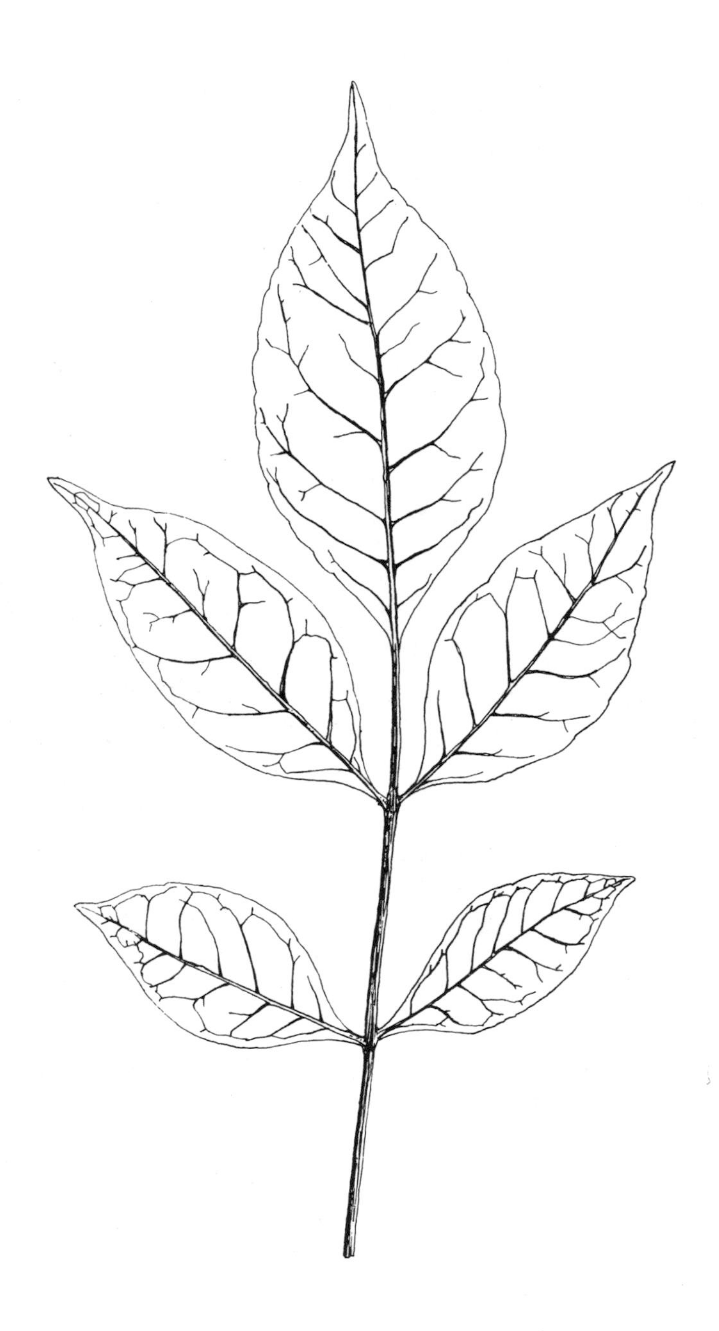

N.T.S. (G.I.B.)

Black Ash

Black Ash

(Brown Ash)

Fraxinus nigra Marsh.

Range: Southern Canada south to central U.S.

Habitat: Swamps and shores.

Profile: Small to medium sized tree. Open, scraggly crown.

Branching: Opposite.

Leading Shoot: Straight.

Leaves: Compound, 12–16" (30–40 cm) long. Leaflets 7–13 in number. NO STALKS *between leaflets and leaf stem.* Margin of *leaflets finely toothed,* surfaces smooth except for tufts of hair on veins on under side. Leaflets narrow, oblong.

Buds: Decidedly black, oval, pointed, as broad as long. First pair of lateral buds *some distance below the terminal.*

Twigs: New twigs smooth. Very stout, light gray. Pith solid yellow or buff-colored.

Bole: Somewhat crooked, with knobby bunches.

Bark: CORKY ON OLDER TREES, EASILY RUBBED OFF WITH THE HAND. Flaky gray outside and light brown underneath.

Flowers: Male and female flowers on the *same tree.* Long panicles appearing before the leaves

Fruit: Wing narrowed at seed cavity, 1–1 1/2" (2–3 cm) long.

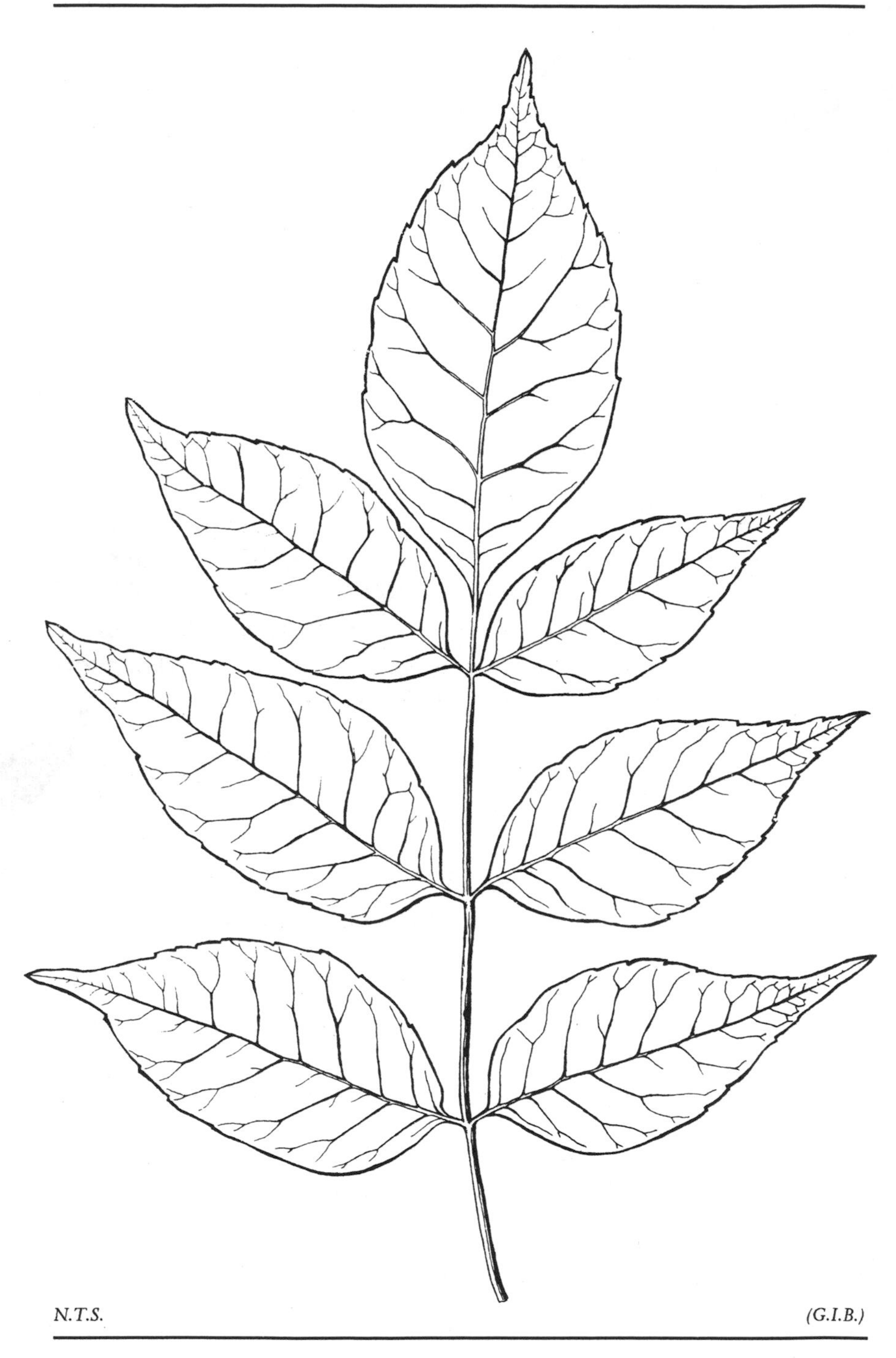

White Ash

White Ash

(American Ash)

Fraxinus americana L.

Range: Southern Canada south to southern U.S.

Habitat: Rich woods.

Profile: Large, tall, straight tree.

Branching: Opposite.

Leading Shoot: Straight.

Leaves: *Compound,* 8–12" (20–30 cm) long with *5–9 leaflets (mostly 7).* Leaflets 5" (10–12 cm) long. Leaflets *stalked,* i.e. *short stalks between each leaflet and the central stem of the leaf.* Margin of leaflets smooth or slightly wavy-toothed. Surfaces smooth. Leaflets rounded oval.

Buds: Rounded, dark brown, rough. Surface hairy. First pair of lateral buds at the same level, *directly below the terminal.*

Twigs: Opposite branched, light greenish brown with small white dots and a slight bloom. Pith solid.

Bole: Straight, free of branches below.

Bark: Smooth when young, grooved in older trees. Longitudinal ridges forming diamond-shaped patterns, separated by narrow interlacing ridges. Firm.

Flowers: Male and female flowers on *separate* trees, small and inconspicuous, appearing with the leaves,

Fruit: Winged, resembling a canoe paddle. 2" (5 cm) long. Not narrowed at seed cavity.

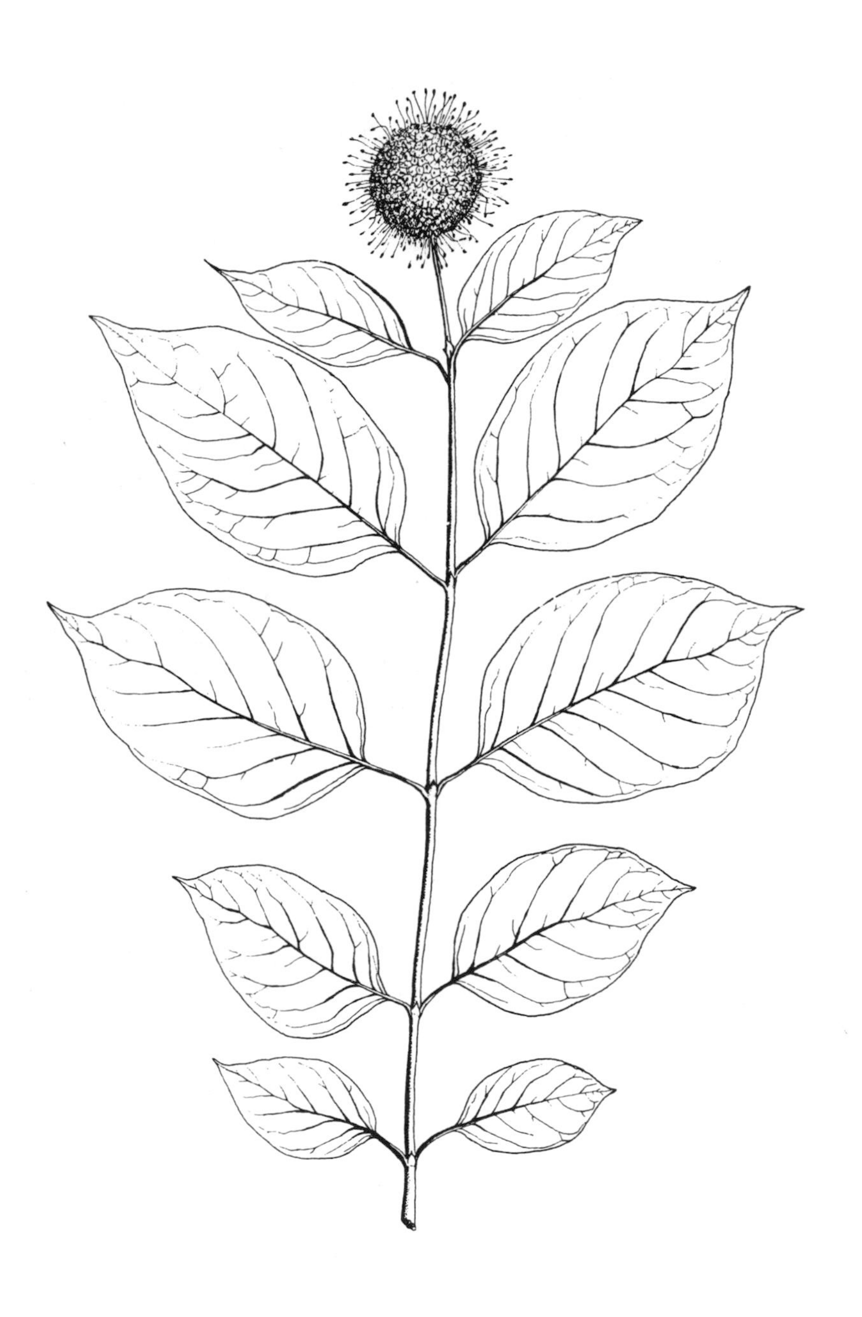

N.T.S. *(G.I.B.)*

Buttonbush

Buttonbush

Cephalanthus occidentalis L.

Range: Nova Scotia to Florida.

Habitat: Wet soil and swamps, lake shores.

Profile: Erect shrub up to 6' (2 m).

Branching: Opposite.

Leaves: Opposite or in whorls of three. 3–6" (7–15 cm), smooth, *no teeth,* pointed at tip, broad at base, dark green, glossy above, slightly downy beneath. Stalks short.

Buds: Small in depressed areas, surrounded by bark.

Twigs: Slender, round or angled, light reddish brown, *pith light brown 4–6 sided.*

Bark: Gray, cracked, flaky.

Flowers: Tubular, creamy white, in dense round heads, crowded in round balls, 1–1½" (2–4 cm) in diameter, long stalked, often in groups of 3. Fragrant, July–August.

Fruit: Small, triangular-shaped capsules, greenish in *ball-like heads* like the flowers. ½–¾" (1–2 cm) in diameter, remaining on the bush in winter.

(G.I.B.)

Bush Honeysuckle

Bush Honeysuckle

Diervilla lonicera Mill.

Range: Canada to Virginia.

Habitat: Dry mountain slopes.

Profile: Low shrub to 3' (1 m).

Branching: Opposite.

Leaves: Smooth, egg-shaped, long pointed. *Sharply toothed,* 2–4" (5–10 cm) long, stalked.

Buds: Narrow, pale brown with 4 or more pointed scales. No terminal bud.

Twigs: Slender, pale brown or straw colored with a hairy *lined ridge.* Scales at twig bases.

Bark: Gray brown and shreddy on old stems.

Flowers: Yellow turning to red, funnel-shaped, 1/4" (.6 cm) in clusters of 2–10 at bases of leaves. Flowers mostly in threes.

Fruit: Very slender, erect pods 1/2" (1 cm) long, a long beaked capsule.

(G.I.B.)

American Fly Honeysuckle

American Fly Honeysuckle

Lonicera canadensis Bartr.

Range: Quebec to North Carolina.

Habitat: Cool, dry rocky woods.

Profile: Low shrub 2–4' (.5–1.5 m).

Branching: Opposite.

Leaves: Smooth both surfaces, broadly and blunt pointed. *No teeth,* but may have a fringe of hairs on the margin, bright green. *Pinnately veined.*

Buds: Olive brown, ovoid, pointed.

Twigs: Slender, smooth, ashy-gray with large pith.

Flowers: Pale greenish yellow, narrowly funnel-shaped. *Borne in pairs* on one stalk, 1" (2.5 cm) long.

Fruit: Long red berries in pairs.

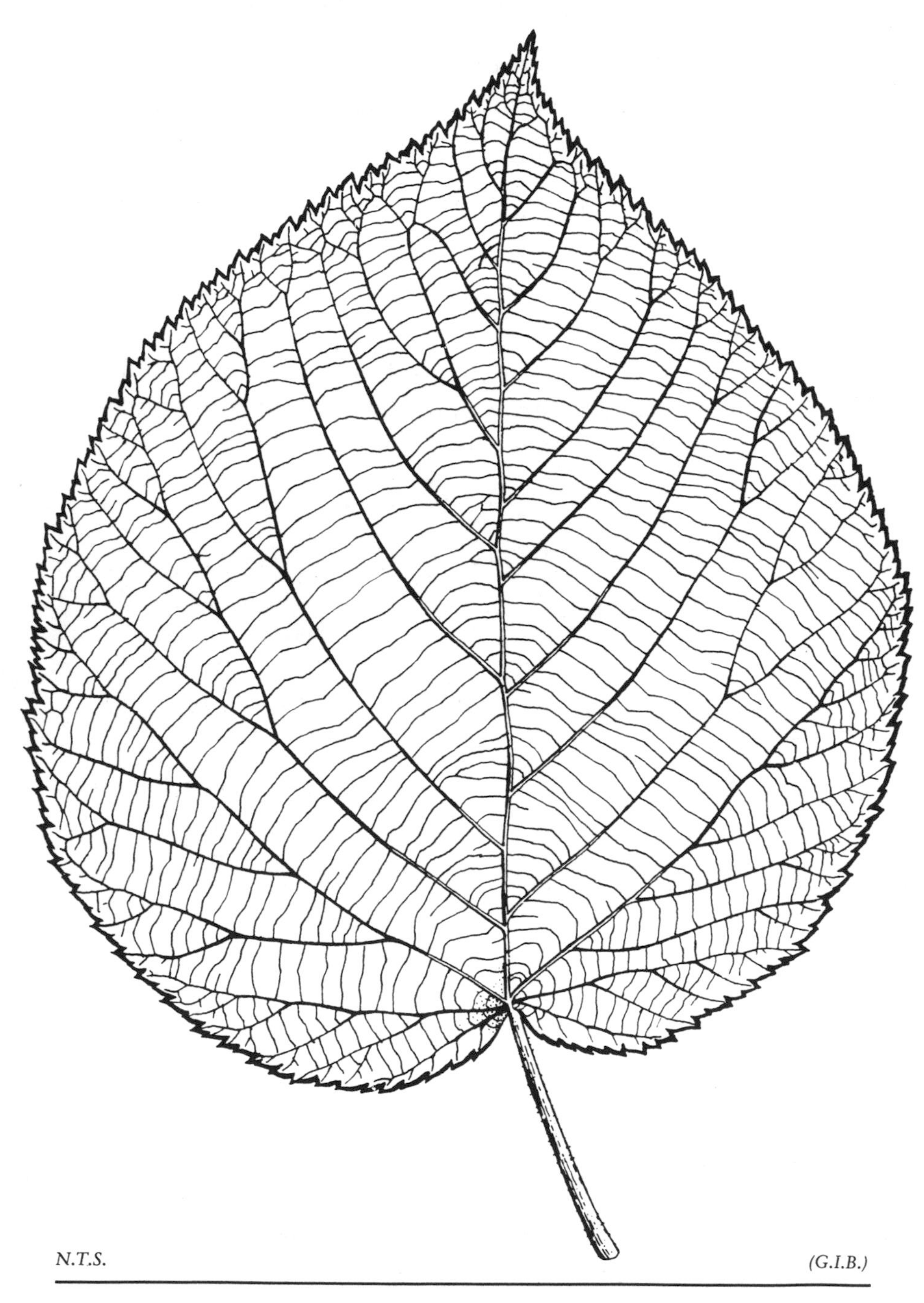

N.T.S. (G.I.B.)

Hobblebush or Witch Hobble

Hobblebush or Witch Hobble

(Moosewood, Tangle-legs)

Viburnum alnifolium Marsh.

Range: Southern Canada to middle Atlantic states and west to Lake states.

Habitat: Cool, moist ravines, shady lake shores.

Profile: Shrub 3–8' (1–2 m) high, spreading.

Branching: Opposite.

Leading Shoot: Straight.

Leaves: *Very large,* 4–8" (10–18 cm) long. *Almost round, finely toothed, heart-shaped at base. Scurfy, hairy beneath,* rusty, veins much branched. Stalks with brownish hairs.

Buds: Large, long, in pairs, light brown covered with *velvety fuzz. Flower buds rounded.* Occur only at tips of twigs.

Twigs: *Tough, stout, olive* brown to light purplish brown. Coated with rusty brown hairs toward the tip.

Branches: Reclining, sometimes taking root.

Bark: Light brown, with wavy grooves.

Flowers: Large white clusters with 5 petals, appearing before leaves are fully grown. May–June.

Fruit: Red berries turning black. Stones with a groove on each edge.

Remarks: Viburnums likely to be confused with dogwoods. Viburnum leaves are nearly always toothed, or lobed, while dogwood leaves are smooth on the margins (entire) and veins are parallel.

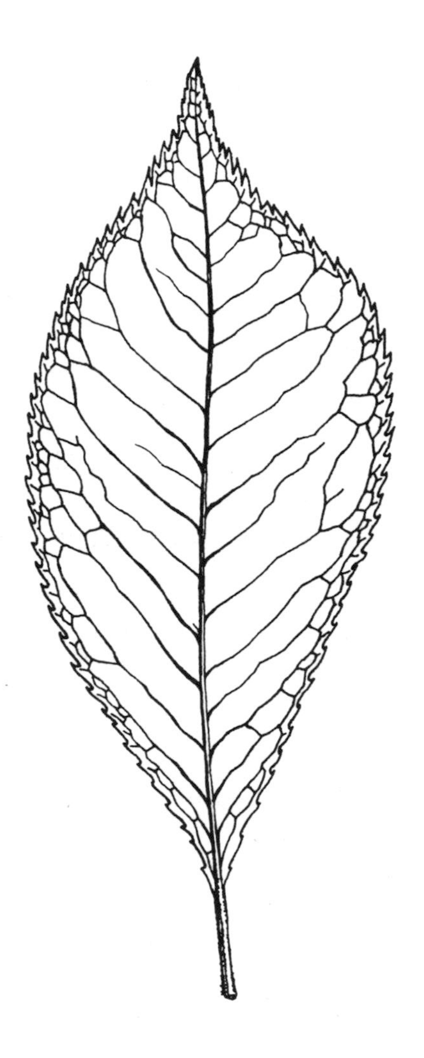

(G.I.B.)

Withe Rod

Withe Rod

(Withe Bush, Wild Raisin)

Viburnum cassinoides L.

Range: Southern Canada to Lake states, south to southern Appalachians.

Habitat: Swamps and streams, borders of woods.

Profile: Erect shrub 3–10' (1–3 m), forming dense thickets.

Branching: Opposite.

Leading Shoot: Straight.

Leaves: *Narrow,* loosely toothed, 2–6" (5–15 cm) long on short stalks, *tapering to an abrupt point.* Dull green above, leathery, teeth inconspicuous, as small notches.

Buds: Covered by single pair scales. Terminal bud long, yellow or golden. 2 light brown scales split curved. Flower buds large and flask-shaped.

Twigs: *Slender, long, flexible,* dull, pale grayish brown. Slightly hairy. Somewhat angled, with indistinct lenticel dots.

Branches: Erect, long, straight.

Bark: Pale brown, smooth.

Flowers: Small, white in flat clusters, ill-scented. Borne on stalks 1" (2.5 cm) long. June.

Fruit: 1/2" (1 cm), white, turning pink and dark blue. Pulp sweet with heavy bloom.

(G.I.B.)

Maple-Leaved Viburnum

Maple-Leaved Viburnum

(Dockmackie)

Viburnum acerifolium L.

Range: Southern Canada to Minnesota, south to Georgia and Tennessee.

Habitat: Dry and rocky woods.

Profile: Slender branched shrub, 2–6' (.5–2 m).

Branching: Opposite.

Leading Shoot: Straight.

Leaves: *Maple-like,* 3-lobed, up to 5" (12 cm), broad *downy beneath* with many dots. Coarsely and *regularly toothed all around,* turning *purple* in autumn.

Buds: Long, often tinged with red. *Smooth, greenish,* 2-scales. Terminal bud larger. Opposite.

Twigs: Slender, round, light grayish brown. Minutely hairy. Large white pith.

Bark: Gray-brown.

Flowers: White clusters on slender stalks, small 1½–3½" (3–8 cm) across, often creamy or pinkish.

Fruit: *Bluish-black* in terminal clusters.

N.T.S. (G.I.B.)

Highbush Cranberry

Highbush Cranberry

Viburnum trilobum Marsh.

Range: Newfoundland to British Columbia south to New England, Pennsylvania, Illinois, South Dakota, Wyoming. Rare in southern New England.

Habitat: Cool woods.

Profile: Medium to tall shrub up to 12' (4 m).

Branching: Opposite.

Leading Shoot: Straight.

Leaves: Maple-like, 3–5 ribs, *few teeth. 3 lobes, long pointed. Basal one-third without teeth.* Teeth indistinct or absent, leaf stalks with round-tipped hair.

Buds: Egg-shaped, chestnut brown, *small hairs.*

Twigs: Slender, smooth, gray or ashy scattered but prominent lenticels (streaks). Large white pith.

Bark: Gray or ashy, prominent lenticels (streaks).

Flowers: Large white clusters, outer rim, enlarged and showy 2–4" (5–10 cm).

Fruit: *Orange to red,* juicy, in terminal flat-topped, drooping clusters.

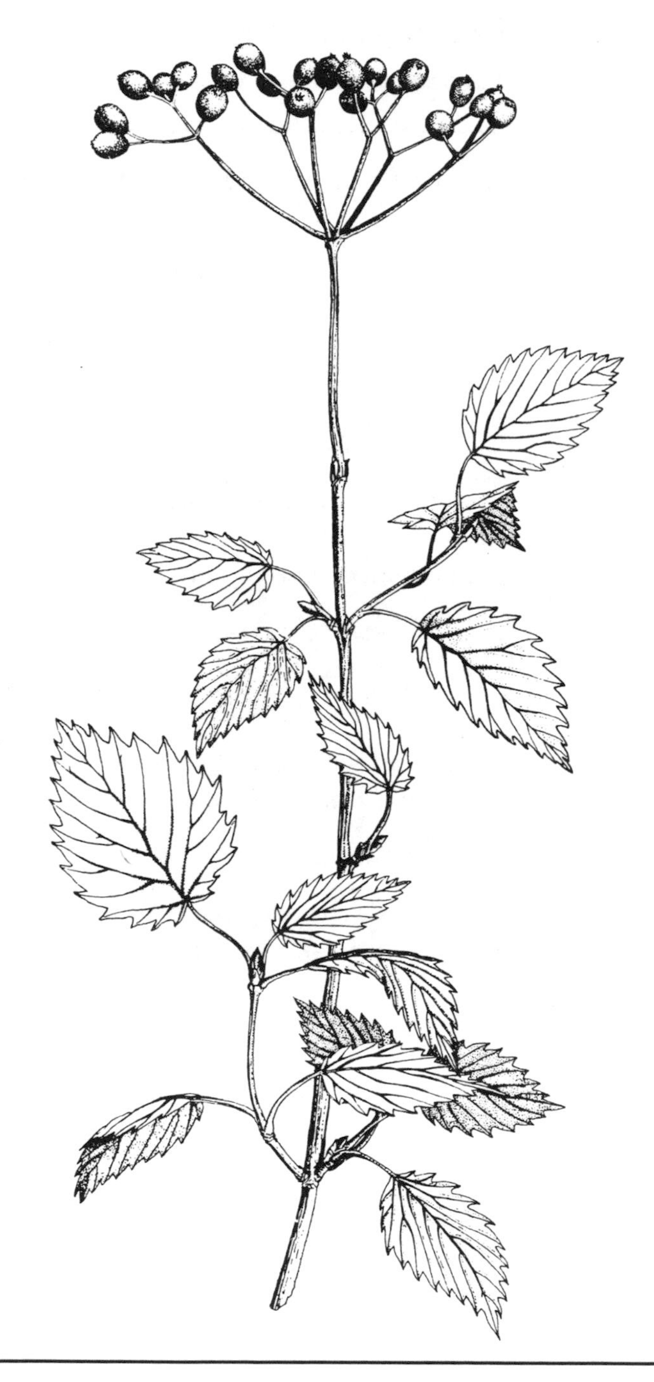

Smooth Arrowwood

Smooth Arrowwood

Viburnum recognitum Fern.

Range: Northern New England, south to southern New Hampshire.

Habitat: Mostly on dry soils or well-drained rich soils.

Profile: Small shrub, erect 3–9' (1–3 m) much branched.

Branching: Opposite.

Leading Shoot: Straight.

Leaves: *Nearly round* with *saw-like coarse teeth,* 1–2" (2–5 cm), smooth.

Buds: Reddish-brown with 2 pairs of scales . *Opposite* pointed, 1/4" (.5 cm) long.

Twigs: Long, straight, slender, arrow-like, smooth.

Flowers: Small, flat clusters, white 2–4" (5–10 cm) across. Appear in June.

Fruit: Blue-black 1/3" (1 cm), August, stone with furrow.

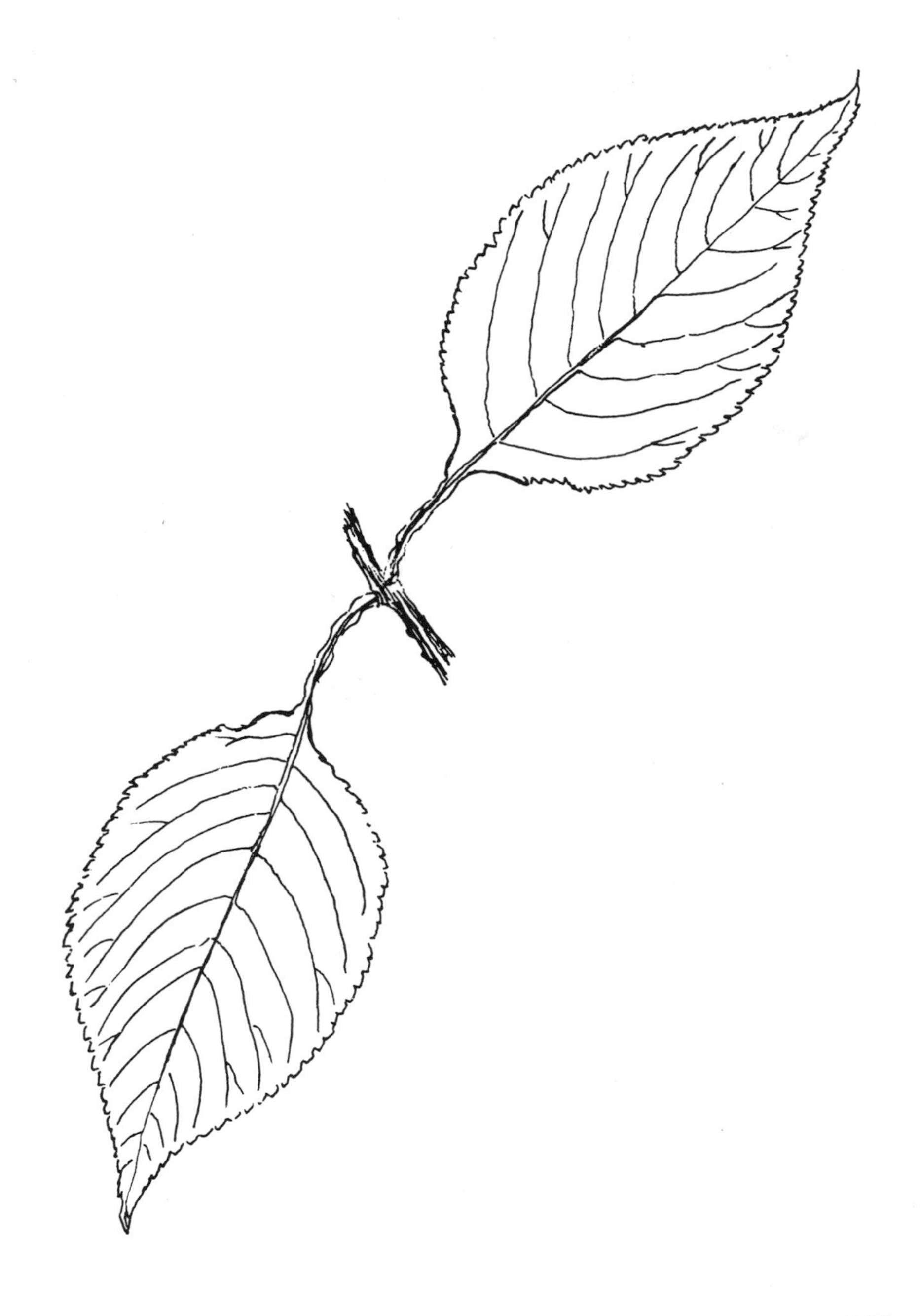

(P.K.)

Nannyberry

Nannyberry

Viburnum lentago L.

Range: Southern New England, north to southern New Hampshire and Lake Champlain.

Habitat: Rich, moist soils.

Profile: Large shrub up to 25' (7 m).

Branching: Opposite.

Leading Shoot: Straight.

Leaves: *Oval,* finely and *evenly toothed. Leaf stalks with wavy margins and wings.*

Buds: Drab terminal bud large 3/4" (2 cm) long, somewhat hairy.

Twigs: Heartwood with nasty odor.

Bark: Slender, flexible, reddish brown to ashy.

Flowers: Clusters, white, *lacking any main stem.* June.

Fruit: Blue-black 1/4–1/2" (.6–1 cm), September, sweet and edible.

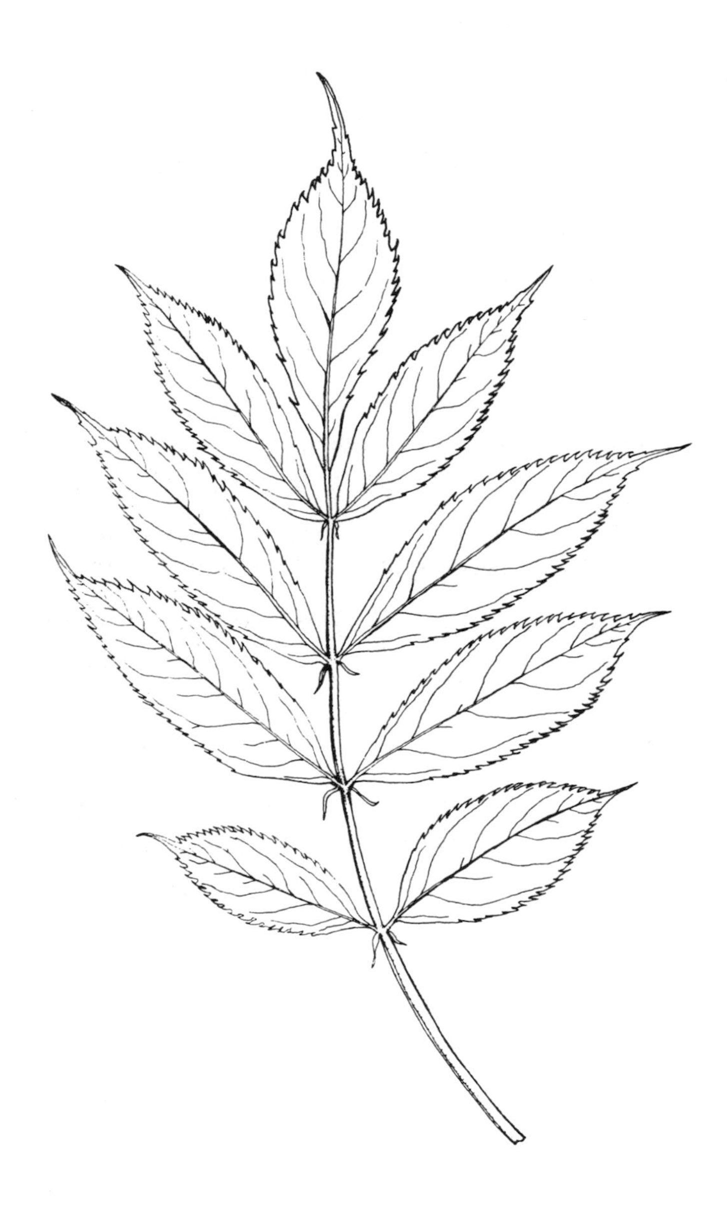

N.T.S. *(G.I.B.)*

Common Elder

Common Elder

(White-Berried Elder)

Sambucus canadensis L.

Range: New England.

Habitat: Rich woods, openings.

Profile: Small shrub, woody to partly herbaceous.

Branching: Opposite.

Leaves: Compound, with 5 to 11 leaflets, usually 7. Sharply toothed. Lower leaflets often 3-parted. Upper surface smooth, lower surface pale and smooth or downy on veins.

Buds: Small, oblong, covered with brownish scales.

Twigs: Pale yellow-brown, smooth, evil-smelling when bruised. Fluted or squared PITH VERY LARGE, WHITE, long internodes.

Bark: Inner bark green.

Flowers: *Flat-topped* clusters of small white flowers, *blooming much later* than red-berried elder, i.e. *late June or July*, odor good.

Fruit: *Purplish black,* berries in flat-topped clusters. Late August to early *October.*

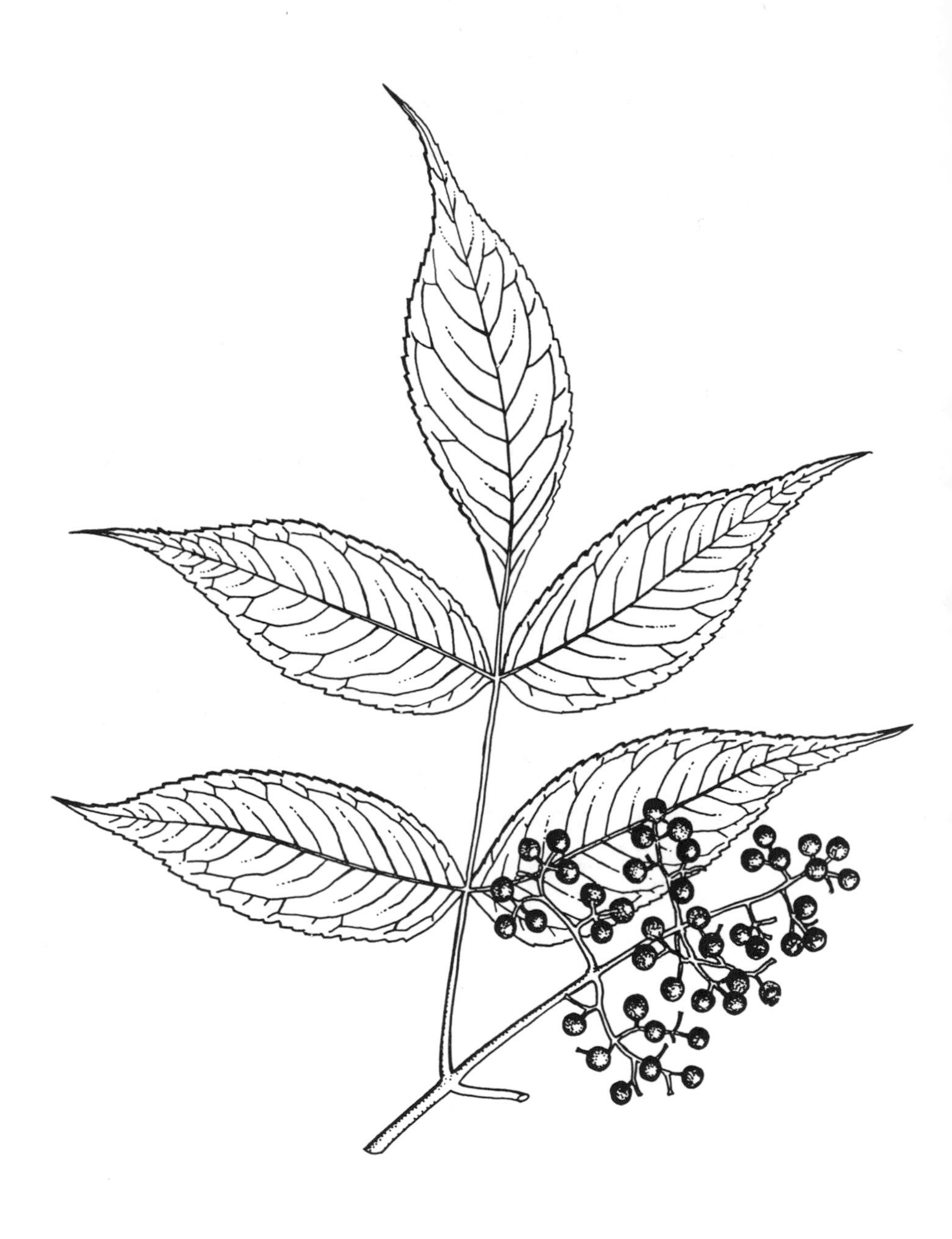

N.T.S. (G.I.B.)

Red-Berried Elder

Red-Berried Elder

Sambucus pubens Mich.

Range: New England.

Habitat: Woods and openings, often rocky.

Profile: Small shrub.

Branching: Opposite .

Leaves: Similar to common elder. Terminal leaflet has a definite stalk.

Buds: Very large, opposite on stem, plump, oval, dark purplish red, no terminal bud.

Twigs: ASHY brown, lenticles conspicuous as raised brown lines, PITH BROWN.

Bark: Warty.

Flowers: Dense *pyramids* of small white flowers, *blooming very early in the spring; April–May.* Heavy odor.

Fruit: *Bright red,* in pointed clusters. Ripe in *June or early July.*

Short Cuts to Identification of Trees & Shrubs

BUDS

Description:	*Species:*
Large, conspicuous at tip of branch	Hickory, butternut. ash
Long, pointed, slender	Beech, birch, shadbush
Only one cap-like scale on bud	Willow
Clustered at ends of branches	Oak, azalea
Many scales	Poplar
No terminal bud	Willow
Without scaly covering	Viburnum, witch hazel

LEAVES (NEEDLE-LIKE)

Description:	*Species:*
Needles coming out in bundles or clusters	Pines, larch
Needles dark green above, yellow below	Yew
Needles aromatic	N. white cedar, white spruce, balsam fir
Needles on short stalks	Spruces
Needles with no stalks, growing directly out of twig	Balsam fir, hemlock
Scale-like	Cedars
Short Needles	Hemlock, spruces, fir
Two kinds of needles on same tree	Red cedar
Clusters of short needles	Larch
Two white bands on under side	Balsam fir, hemlock
Clusters of needles growing out of the bark	Pitch pine

LEAVES (BROAD-LEAVED)

Description:	*Species:*
No teeth on margin	Dogwoods
Irregular or lopsided shape	Witch hazel, sassafras, basswood
Small, sharp teeth on margin	Cherries, birches, hornbeam
Large lobes	Maples, oaks
Round or nearly round	Basswood, hobblebush
Leaf stem (petiole) flattened	Aspens, poplars

TWIGS AND BRANCHES

Description:	*Species:*
Aromatic odor or taste	Black birch, yellow birch, sweet gale, sweet fern, spice bush, sassafras
Rancid odor or taste	Cherries
Thorns or prickles	Black locust, hawthorn, Apple, roses, blackberries, gooseberries
Unusual color	
Green:	Yew, white pine, striped maple

TRUNKS COLORED

Description:	*Species:*
White	Paper birch, gray birch
White streaks	Striped maple
Curly yellow	Yellow birch
Reddish	Young paper birch

PITH

Description:	*Species:*
Colored	
Orange:	Red-berried elder
Brown:	Black locust, sumac
Chambered	Butternut, black gum

Other Books on Tree Identification

Archibald, David. *Trees of Northeastern and Central North America.* Doubleday and Co., Garden City, N.Y., 1967. Quick Key Guide Series.

Bernath, Stefen. *Trees of the Northeast Coloring Book.* Dover Publications, 1979.

Blakeslee, A. F. and C. D. Jarvis. *New England Trees in Winter.* Reprinted (originally issued as Bull. #69), Storrs Agric. Exp. Sta., Storrs, Conn., 1911.

Dayton, William A. *United States Tree Books.* Bibliographical Bull. 20, U. S. Dept. of Agric., Washington, D.C., 1952. A list of publications on tree identification.

Fernald, M. L. *Gray's Manual of Botany, 8th Ed.* Am. Book Co., 1950.

Gleason, H.A. and A. R. Cronquist. *Manual of Vascular Plants of the Northeastern States and Adjacent Canada.* D. Van Nosftrand Co., Princeton, N.J., 1963, 810 pp.

Graves, Arthur Harmount. *Illustrated Guide to Trees and Shrubs, 1952.* Privately printed; Reprinted by Harper and Row, 1956.

Grimm, W. C. *The Book of Shrubs.* Bonanza Books, N.Y., 1957.

Harlow, W. M. and E. S. Harrar. *Textbook of Dendrology.* McGraw Hill, 1937.

Hodgdon, A. R. and G. M. Moore. *Summer Key to Trees of New Hampshire.* Society for the Protection of New Hampshire Forests, 1961.

Important Trees of Eastern Forests. U. S. Forest Service, 1968. 111 pp.

Knobel, Edward. *Identify Trees and Shrubs by Their Leaves.* Dover Publications, N.Y.

Petrides, George A. *Field Book to Trees and Shrubs.* Peterson's Field Guide Services, 2nd Ed. Houghton Mifflin Books, 1958. 42 pp.

Rogers, Matilda. *A First Book of Tree Identification.* Random House, N.Y., 1951. 95 pp.

Sargent, Charles Sprague. *The Solva of North America.* Houghton-Mifflin Co., N.Y., 1902. 22 tomes.

Schuyler, F. Matthews. *Field Book of American Trees and Shrubs of Northern New England.* Putnams, 1915.

Steele, Frederic L. *Beginner's Guide to Trees and Shrubs of Northern New England.* Originally printed 1971, reprinted 1979. Society for the Protection of New Hampshire Forests. 35 pp. $1.25.

Steele, Frederic L. and Albion R. Hodgdon. *Trees and Shrubs of Northern New England.* Society for the Proteccion of New Hampshire Forests, 1968. 127 pp. $2.75.

Sargent, C. S. *Manual of the Trees of North America.* Reprinted in 2 Vol. by Dover Publications.

The New York Times Book of Trees and Shrubs. 1964. 118 pp.

Index

Henry Ives Baldwin, 1896–1992, was a pioneering and internationally-known research forester and botanist. Research Forester for the State of New Hampshire from 1931 to 1962, he was an authority on forest genetics and tree seed biology. His long life of work in the woods made him a consummate observer of the distinguishing characteristics of trees and shrubs.

Gunnar Ives Baldwin, Jr., a grandson of the author, is a professional landscape painter. He is a graduate of Yale in fine arts. He resides in Plymouth, New Hampshire.

Priscilla Kunhardt studied at the Art Students League in New York City. She has contributed many drawings to the New England Wild Flower Society and the Audubon Society of New Hampshire.